GUIDE
TO THE USE OF THE
WIND LOAD PROVISIONS
OF ASCE 7-95

by Kishor C. Mehta
Texas Tech University

Richard D. Marshall
Wind Research Services

Published by ASCE PRESS

American Society of Civil Engineers
1801 Alexander Bell Drive
Reston, Virginia 20191-4400

Abstract: The objective of the *Guide to the Use of the Wind Load Provisions of ASCE 7-95* is to provide guidance in the use of the wind load provisions set forth in ASCE Standard 7-95. The Guide is a completely new document because the wind load provisions underwent major changes from the previous ASCE Standard 7-88 (or ASCE 7-93). The Guide contains six example problems, worked out in detail, which can provide direction to practicing professionals in assessing wind loads on a variety of buildings and other structures. Errata and clarifications from the previous Guide are also included.

Library of Congress Cataloging-in-Publication Data

Mehta, Kishor C.
Guide to the use of the wind load provisions of ASCE 7-95 / by Kishor C. Mehta, Richard D. Marshall.
p. cm.
Includes bibliographical references and index.
ISBN 0-7844-0302-3 (alk. paper)
1. Wind–pressure. 2. Structural engineering. 3. Building–Standards–United States. I. Marshall, Richard D. II. Title.
TH891.M45 1997 97-45835
624.1'72'021873–dc21 CIP

Library of Congress Catalog Card No: 97-45835
ISBN 0-7844-0302-3
Manufactured in the United States of America.

CONTENTS

LIST OF TABLES

LIST OF FIGURES

Preface

This guide is designed to assist professionals in the use of the wind load provisions of the American Society of Civil Engineers Standard ASCE 7-95. The ***Guide*** is a completely new document because the wind load provisions underwent major changes from the previous Standard ASCE 7-88 (or ASCE 7-93). The ***Guide*** contains six example problems worked out in detail which can provide direction to practicing professionals in assessing wind loads on a variety of buildings and other structures. Every effort has been made to make these illustrative example problems correct and accurate. The authors would be pleased to receive comments regarding inaccuracies, errors or different interpretations. The views expressed and interpretation of the wind load provisions made in the ***Guide*** are those of the authors and not the ASCE 7 Standards Committee or the ASCE organization.

Authors' Disclaimer

Although the authors have done their best to ensure that any advice, recommendation, interpretation or information given herein is accurate, no liability or responsibility of any kind (including liability for negligence) is accepted by the authors.

Acknowledgments

The authors wish to acknowledge the members of the ASCE 7 Standards Committee that was chaired by Dr. Walter P. Moore, Jr. during the consensus process of ASCE 7-95. The members of the Wind Load Task Committee and the ASCE 7 Standards Committee contributed significantly to the final wind load provisions of ASCE 7-95 through their questions, comments and discussions. The authors are indebted to these one hundred plus members.

During its development the Guide was reviewed critically by Dr. Dale C. Perry, Texas A&M University and Dr. James R. McDonald, Texas Tech University. They particularly provided input to the example problems and checked them for accuracy. Discerning input by these eminent wind engineers is appreciated by the authors.

In a document of this type there are individuals in the background who help in layout, word processing and checking calculations. These tasks were handled by the staff of the Wind Engineering Research Center, Texas Tech University. One graduate student, John McAnulty, performed exceptionally in checking the numbers in the worked examples. The authors are indebted to all of these individuals for their commendable efforts.

Chapter 1.
INTRODUCTION

The nature of wind and its effects has taken an important role in the design of buildings and other structures to prevent loss of lives and to mitigate property losses. Hurricane Hugo in 1989 which affected Puerto Rico and South Carolina, Hurricane Iniki in 1992 which struck Hawaii and Hurricane Andrew in 1992 which devastated South Florida have indicated the importance of wind loads in design and construction of ordinary low buildings. Research and development in assessing wind loads for buildings and structures is continuing. Research is being pursued in many areas including probabilistic assessment of wind speeds, translation of wind speeds into loads through wind tunnel testing, validation of this translation through full-scale testing, documentation of damage and economic assessment of construction. The goal is to establish wind load designs that prevent injuries and fatalities and reduce property damage to an acceptable level.

The American National Standards Institute published a revision of ANSI A58.1 - 1972 in 1982 (ANSI, 1982). This standard, *Minimum Design Loads for Buildings and Other Structures*, contained an innovative approach to wind loads for components and cladding of buildings. Wind load specification was based on understanding the aerodynamics of wind pressures in building corners, eaves and ridge areas, as well as the effects on pressures of area averaging.

In the mid 1980's the American Society of Civil Engineers (ASCE) assumed responsibility for the committee which establishes design loads for buildings and other structures. The ASCE Committee 7 reviewed the wind load provisions of ANSI A58.1-1982 and made minor changes and clarifications which were incorporated in *Minimum Design Loads for Buildings and Other Structures*, ASCE 7-88 (ASCE, 1990). A revised version of ASCE 7-88 was published as ASCE-93. However, the wind load provisions were not changed. Thus, the wind load provisions have remained essential the same as ANSI A58.1-1982 until the publication of ASCE 7-95.

Members of the Task Committee on Wind Loads, listed in Table 1, developed the wind load provisions of ASCE 7-95 through the consensus process. The ASCE 7 Committee on Minimum Design Loads, consisting of more than 90 members, voted on and approved the new wind load provisions after eight committee ballots and two public ballots. The consensus process required that every negative vote be resolved through written response and that 75% of the returned ballots be affirmative.

There are several new items added in the ASCE 7-95 wind load provisions, including a topography factor, alternate pressure coefficients for main wind-force resisting systems of low-rise buildings, pressure coefficients for additional roof geometries, and provisions for full and partial loading of high-rise buildings. These new items are discussed in Chapter 4. A major change in the wind load provisions is the use of a 3-second gust as the basic wind speed instead of the fastest-mile wind speed. The 3-second gust speed required modification of exposure coefficients, gust effect factors, and pressure coefficients for components and cladding.

The main reason for adopting the 3-second gust speed as the reference wind speed is that the National Weather Service no longer collects fastest-mile wind speed data. Since assessment of wind speed on a probabilistic basis depends on historical records of wind speed, it reasons that gust speed will provide a good reference wind speed in the future. Other reasons are: (1) 3-second gust speed data are collected at a large number of stations around the country, and (2) gust speed is consistent with terminology used by meteorologists, the news media and the public. Development of the new wind speed map is discussed in Chapter 4.

Table 1
Membership of the ASCE 7 Wind Load Task Committee (1989 - 1995)

Committee Member	Affiliation	Location
Dr. Kishor C. Mehta, Chairman	Texas Tech University	Lubbock, TX
Dr. Jack E. Cermak	Colorado State University	Fort Collins, CO
Dr. Ahsan Kareem	University of Notre Dame	Notre Dame, IN
Mr. Gill Harris	CECO Building Systems	Columbus, MS
Dr. Richard D. Marshall*	Wind Research Services	Poolesville, MD
Dr. Dale C. Perry	Texas A&M University	College Station, TX
Mr. Herb Saffir	Herbert S. Saffir Consulting Engineers	Coral Gables, FL
Mr. Thomas Smith	National Roofing Contractors Association	Rosemont, IL
Dr. Ted Stathopoulos	Concordia University	Montreal, Canada
Mr. Rick Vognild	Southern Building Code Congress International	Birmingham, AL

* Formerly with the National Institute of Standards and Technology, Gaithersburg, MD

1.1 OBJECTIVE OF THE GUIDE

The objective of this Guide is to provide direction in the use of the wind load provisions of ASCE 7-95 (referred to as the Standard). Background material which forms the basis for the Standard is reviewed in order to understand the provisions and their limitations. Additional background material is given in the Commentary of the Standard; this material is not repeated here. In-depth background material can be obtained from the cited references.

This Guide contains six worked examples. Sufficient details of calculations of wind loads are provided to help the reader properly interpret the wind load provisions of the Standard. Section 6 of the Standard, as well as the figures and tables of the Standard, are cited liberally in this Guide and the example problems as needed. To avoid repeating the words *ASCE 7-95* or *the Standard,* the section, table and figure numbers are given without additional reference. It is necessary to have a copy of ASCE 7-95 to follow the examples and work with this Guide.

1.2 SIGNIFICANT CHANGES

The most significant change in ASCE 7-95 is the use of 3-second gust speed instead of fastest-mile wind speeds. This change necessitated revision of terrain and height factors, gust effect factors and pressure coefficients for components and cladding.

Additional significant changes and additions are listed below:

- Provisions have been added for wind speed-up over isolated hills and escarpments (topographic effect).
- New provisions have been added for full and partial loading on main wind-force resisting system of buildings with a mean roof height greater than 60 ft (torsional loading effect).
- An alternate procedure has been added for determining external loads on main wind-force resisting systems of low-rise buildings.
- Internal pressure coefficients have been increased for partially enclosed buildings and for buildings located in hurricane-prone regions.
- Pressure coefficients for components and cladding have been added for hipped, stepped, multispan gable, monoslope and sawtooth roofs.
- Velocity pressure exposure coefficients have been revised to be compatible with 3-second gust speed.
- Gust effect factor procedures are unified for flexible and non-flexible buildings and structures. A new procedure is provided for calculating a gust effect factor.
- Pressures for components and cladding of buildings with a mean roof height less than 60 ft are reduced for buildings sited in Exposure B.

Other changes in tables and figures and in the wind tunnel test procedure incorporate the latest available technical information. The basic format of the Standard remains the same as in ASCE 7-88. Background on some changes is given in Chapter 4. Additional information can be obtained from the Commentary of the Standard and from the references.

1.3 LIMITATIONS

There are limitations on the analytical procedure given in the Standard. These limitations primarily relate to the estimation of basic wind speed, gust effect factor, and shape-dependent pressure coefficients.

Estimation of basic wind speed depends on the quality and length of wind records and the statistical model used. Calibration of the anemometer, accurate records of the instrument height and appropriate assessment of surrounding terrain roughness can affect the quality of the wind speed data. In addition, a longer data set provides a more reliable statistical estimation. In the development of the wind speed map of Figure 6.1, a new procedure was introduced whereby data from several stations were combined, when found independent, to obtain a longer data set and thus reduce sampling error. If an individual station is used to assess basic wind speed, as is permitted by Section 6.5.2.2, the sampling error at that location will be larger than that associated with the wind speed map.

Several special wind regions are indicated on the wind speed map of Figure 6.1. These regions include mountain ranges, gorges or river valleys. Unusual winds may be encountered in these regions because of orographic effects or because of the channeling of wind. Consultation with a wind engineer, a meteorologist, or a building official is advised to assess any anomalies in the wind climate.

Tornado winds are not included in development of the basic wind speed map because of their rare occurrence. Intense tornadoes can have ground level wind speeds in the range of 150-200 mph; however the annual probability of occurrence of this range of wind speeds may be less than 1×10^{-5} (mean recurrence interval exceeding 100,000 years). Special structures and storm shelters can be designed to resist tornado winds if required.

Pressure and force coefficients are obtained from wind tunnel testing and, to a limited extent, from full-scale testing. The Standard contains pressure and force coefficients for the most common building shapes. These coefficients were obtained from atmospheric boundary-layer wind tunnel tests or have been used successfully in practice for a long time.

It is possible to use pressure and force coefficients from other standards and reports provided their definition can be made consistent with the 3-second gust speed. If different values of pressure coefficients for the same shape are found in the literature, it is recommended that the values in the Standard be used because they have been carefully reviewed through the consensus process.

1.4 TECHNICAL LITERATURE

There has been a vast amount of literature published on wind engineering during the past three decades. Most of it is in the form of research papers in the *Journal of Wind Engineering and Industrial Aerodynamics*, *Journal of Structural Engineering, Journal of the American Society of Civil Engineers*, *Proceedings of the International Conferences on Wind Engineering* (a total of nine), *Proceedings of the U.S. National Conferences on Wind Engineering* (a total of eight), *Proceedings of the U.S. Asia-Pacific Conferences on Wind Engineering* (a total of four) and *Proceedings of the European-African Conferences on Wind Engineering* (two). The literature is extensive and scholarly; however, it is not always in a format that can be used by practicing professionals.

Several textbooks, handbooks, standards and codes, reports and papers contain material that can be used to determine wind loads. Selected items are identified in Table 1.2. The items are listed by subject matter for easy identification. Detailed references for these items are given in References Section citations.

Table 1.2 Technical Literature

Subjects	Selected Reference Material
Wind Effects on Buildings and Structures	Simiu and Scanlan (1996), Lawson, vols. 1 and 2 (1980), Cook, vols. 1 and 2 (1985) Liu (1991), Dyrbye and Hansen (1996), Newberry and Eaton (1974), Holmes and Melbourne (1990)
Foreign Codes and Standards	NRCC (1995a, 1995b), Australian Standard AS1170.2 (1989), British Standard BS6399 (1995), Eurocode Draft (1994)
Wind Tunnel Testing	Reinhold (1982), ASCE (1987)
General Wind Research	Davenport (1979,1978), ASCE (1961), Cermak (1977)
Pressure and Force Coefficients	ASCE (1961), Hoerner (1965)
Tornadoes	FEMA 83-A (1980), Mehta and Minor (1977), McDonald (1983)

Chapter 2.
WIND LOAD PROVISIONS

Wind load provisions specified in ASCE 7-95 follow the same format as ASCE 7-88, ASCE 7-93 and ANSI A58.1-1982. In this chapter, the format and analytical procedure are presented. In addition, equations for the graphs in Figures 6-5 through 6-8 from Section 6 of the Standard are given since these graphs as presented in the Standard are difficult to interpolate. In order to avoid repeating the words *ASCE 7-95* or *the Standard*, the section, table and figure numbers are given without any additional reference. A few inadvertent errors and ambiguities in ASCE 7-95 that have surfaced are clarified in this chapter.

2.1 FORMAT

The provisions of the Standard permit the determination of wind loads on buildings and other structures using either analytical or wind tunnel procedures. Criteria for credible wind tunnel testing are specified in the Standard. The analytical procedure requires the determination of wind loads on main wind-force resisting system (MWFRS) and components and cladding (C&C) separately. MWFRS is defined in Section 6.2 as the overall structure receiving wind loading from more than one surface. Cladding receives wind loads directly and generally transfers the load to other components or to the MWFRS.

Equations for the determination of wind loads using analytical procedures are given in Table 6-1 of the Standard. Table 6-1 is divided into several sections based on type and height of structure. The Standard recognizes buildings and other structures, flexible buildings and other structures, and buildings with mean roof height less than 60 ft and greater than 60 ft. Answers to the following questions will refer to the proper part of the Table.

1. Is the building or structure flexible or non-flexible (See Section 6.2 for definition)? If flexible, use the bottom half of Table 6.1, otherwise use the upper half of the table.
2. Is the building enclosed or open?
3. Is the mean roof height of an enclosed building less than or equal to 60 ft? If the answer is *yes*, MWFRS wind loads are obtained using coefficients from Figure 6-3 or from Figure 6-4, if mean roof height is less than the building width. If mean roof height is greater than 60ft, the equations for MWFRS and for C&C are clearly indicated in Table 6-1.

2.2 ANALYTICAL PROCEDURE

Design wind pressures for buildings and design forces for other structures are determined using the following two-step procedure:

1. Determine velocity pressures q_z at several heights and q_h at mean roof height, and
2. Determine design pressures using equations given in Table 6-1.

Definitions of the terms of the equations are given in this section. Discussion and background of each term are provided in Chapter 4.

Velocity Pressure, q

Velocity pressure, q, at any height above ground and at mean roof height is obtained for the equation:

$$q_z = 0.00256 K_z K_{zt} V^2 I \qquad \text{(Eq. 6-1)}$$

where q: effective velocity pressure to be used in equations of Table 6-1; q_z is based on K_z at any height, z above ground; q_h is based on K_h at mean roof height, h.

K_z: exposure velocity pressure coefficient which reflects change in windspeed with height and terrain roughness, Table 6-3.

K_{zt}: topographic factor which accounts for wind speed-up over hills and escarpments, Figure 6-2 and $K_{zt} = (1 + K_1 K_2 K_3)^2$.

V: basic wind speed which is the 3-second gust speed at 33 ft above ground for exposure category C and is associated with an annual probability of 0.02 (50-year mean recurrence interval), Figure 6-1.

I: importance factor which adjusts wind speed associated with annual probability of 0.02 (50-year mean recurrence interval), Table 6-2.

Design Pressure, p and Forces, F

Equations for the determination of design pressures for buildings and design forces for other structures are given in Table 6-1. The bottom half of Table 6-1 relates to flexible buildings and other structures. Slenderness of the structure can cause along-wind dynamic resonance effects and increased loading. In a vast majority of buildings resonance effects are negligible since the fundamental frequency of vibration of the frame is likely to be greater than one cycle per second (1 Hz). It is judged that when the ratio of building height to least horizontal dimension is less than 4, the fundamental frequency of vibration of the frame will be greater than 1 Hz. Equations and terms for the top half of Table 6-1 are explained here; equations in the bottom half are similar except for resonance effects.

Design pressures for main wind-force resisting systems (MWFRS) and for components and cladding (C&C) are determined separately. Generally, C&C design pressures will be higher because of localized high pressures acting over small areas. MWFRS receive wind pressures from several surfaces, hence the pressures are likely to be smaller than those for C&C.

Calculation of design pressures requires selection of appropriate gust effect factors and pressure or force coefficients. Equations in Table 6-1 guide the user in the selection of appropriate factors and coefficients. Various gust effect factors and pressure and force coefficients specified in the Standard are as follows:

G: gust effect factor for MWFRS of buildings (all heights) and for other structures, Section 6.6.1.

G_f: gust effect factor for MWFRS of flexible buildings and other structures obtained using a rational analysis, Section 6.6.2.
C_p: external pressure coefficients for MWFRS of buildings (all heights), Figure 6-3.
C_f: force coefficients for open buildings and other structures, Tables 6-6 through 6-10.
(GC_{pf}): external pressure coefficients for MWFRS of low-rise buildings, Figure 6-4.
(GC_p): external pressure coefficients for C&C of buildings, Figures 6-5 through 6-8.
(GC_{pi}): internal pressure coefficients for MWFRS and C&C of buildings, Table 6-4.

Sign convention in the Standard is a follows:
+ (plus sign) means pressure acting toward the surface
- (minus sign) means pressure acting away from the surface
Whenever the sign of ± is specified, both positive and negative values should be used to obtain design loads. Values of external and internal pressures are to be combined algebraically to obtain the most critical load.

Design pressures for MWFRS: low-rise buildings
A low-rise building is a building with mean roof height less than or equal to 60 ft and with ratio of mean roof height to least horizontal dimension of less than 1.

$$p = q_h [(GC_{pf}) - (GC_{pi})]$$

where q_h: velocity pressure at mean roof height using exposure category C, irrespective of actual surrounding terrain
(GC_{pf}): external pressure coefficients given in Fig. 6-4; both loading cases A and B at each building corner must be considered for the design of framing (see Example 4 in Chapter 3).
(GC_{pi}): internal pressure coefficients given in Table 6-4

Design pressures for MWFRS: buildings of all heights

$$p = qGC_p - q_h (GC_{pi})$$

where q: q_z or q_h as specified in Figure 6-3.
G: gust effect factor given in Section 6.6.1.
C_p: external pressure coefficients given in Figure 6-3
(GC_{pi}): internal pressure coefficients given in Table 6-4

For flexible buildings, the equation in the bottom part of Table 6-1 should be used and the gust effect factor G_f should be determined using rational analysis.

Design pressures for C&C: buildings with h ≤ 60ft

$$p = q_h [(GC_p) - (GC_{pi})]$$

where q_h: velocity pressure at mean roof height using exposure category C, irrespective of actual surrounding terrain
(GC_p): external pressure coefficients given in Figures 6-5, 6-6, and 6-7
(GC_{pi}): internal pressure coefficients given in Table 6-4

Design pressures for C&C: buildings with h > 60ft

$$p = q\,[(GC_p) - (GC_{pi})]$$

Where q: velocity pressure; q_z for positive pressures and q_h for negative pressures
(GC_p): external pressure coefficients given in Figure 6-8
(GC_{pi}): internal pressure coefficients given in Table 6-4

Note 6 in Figure 6-8 permits use of external pressure coefficients of Figure 6-5B for roofs when $\theta > 10^0$. This is valid even though the last note in Table 6-1 requires that for buildings with mean roof height between 60 ft and 90 ft, the height to width ratio must be less than 1 in order to use Figure 6-5. The note in Figure 6-8 supersedes the note in Table 6-1 for this case since note 6 in Figure 6-8 is valid for all building heights above 60 ft.

Design forces for MWFRS: open buildings and other structures

$$F = q_z\,GC_f\,A_f$$

where q_z: velocity pressure at height z above ground using appropriate exposure category
G: gust effect factor given in Section 6.6.1
C_f: force coefficients given in Tables 6-6 through 6-10
A_f: projected area normal to wind (except for monoslope roofs where coefficients given in Table 6-6 are for pressures acting normal to the roof surface—in this case use surface area)

For flexible structures the gust effect factor G_f should be determined using rational analysis.

Design forces for C&C: open buildings and other structures

The equation and calculations are the same as for the MWFRS. Appropriate force coefficients for individual components or different shapes should be used.

Design pressures and forces for flexible buildings and other structures

Equations for design pressures and forces for flexible buildings and other structures are given in the bottom half of Table 6-1. The equations and terms are similar to the ones in the top half of Table 6-1. The only exception is the gust effect factor, G_f. This G_f should be determined using any rational procedure. One rational procedure is give in the Commentary of the Standard (see Example 6 in Chapter 3).

2.3 EQUATIONS FOR GRAPHS

Figures 6-5 through 6-8 give pressure coefficient values for C&C for enclosed buildings as a function of effective area of each component and cladding. Wind tunnel results found this relationship between pressure coefficient and effective area to be a logarithmic function. The scale of effective area in the figures is log scale which makes it very difficult to interpolate. Equations for each of the lines in these figures are given below. The equations can be used to develop a computer program to determine wind loads.

Figure 6-5A. Walls for Buildings with h ≤ 60 ft

Positive: Zone 4 & 5

$(GC_p) = 1.0$	for $A \leq 10$ sq. ft
$(GC_p) = 1.1766 - 0.1766 \log A$	for $10 < A \leq 500$ sq. ft
$(GC_p) = 0.7$	for $A > 500$ sq. ft

Negative: Zone 4

$(GC_p) = -1.1$	for $A \leq 10$ sq. ft
$(GC_p) = -1.2766 + 0.1766 \log A$	for $10 < A \leq 500$ sq. ft
$(GC_p) = -0.8$	for $A > 500$ sq. ft

Negative: Zone 5

$(GC_p) = -1.4$	for $A \leq 10$ sq. ft
$(GC_p) = -1.7532 + 0.3532 \log A$	for $10 < A \leq 500$ sq. ft
$(GC_p) = -0.8$	for $A > 500$ sq. ft

Figure 6-5B. Gabled and Hipped Roofs

Roofs $\theta \leq 10°$

Positive with and without overhang: Zone 1,2,3

$(GC_p) = 0.3$	for $A \leq 10$ sq. ft
$(GC_p) = 0.4000 - 0.1000 \log A$	for $10 < A \leq 100$ sq. ft
$(GC_p) = 0.2$	for $A > 100$ sq. ft

Negative without overhang: Zone 1

$(GC_p) = -1.0$	for $A \leq 10$ sq. ft
$(GC_p) = -1.1000 + 0.1000 \log A$	for $10 < A \leq 100$ sq. ft
$(GC_p) = -0.9$	for $A > 100$ sq. ft

Negative without overhang: Zone 2

$(GC_p) = -1.8$	for $A \leq 10$ sq. ft
$(GC_p) = -2.5000 + 0.7000 \log A$	for $10 < A \leq 100$ sq. ft
$(GC_p) = -1.1$	for $A > 100$ sq. ft

Negative without overhang: Zone 3

$(GC_p) = -2.8$	for $A \leq 10$ sq. ft
$(GC_p) = -4.5000 + 1.7000 \log A$	for $10 < A \leq 100$ sq. ft
$(GC_p) = -1.1$	for $A > 100$ sq. ft

Negative with overhang: Zone 1 & 2

$(GC_p) = -1.7$	for $A \leq 10$ sq. ft
$(GC_p) = -1.8000 + 0.1000 \log A$	for $10 < A \leq 100$ sq. ft
$(GC_p) = -3.0307 + 0.7153 \log A$	for $100 < A \leq 500$ sq. ft
$(GC_p) = -1.1$	for $A > 500$ sq. ft

Negative with overhang: Zone 3

$(GC_p) = -2.8$	for $A \leq 10$ sq. ft
$(GC_p) = -4.8000 + 2.0000 \log A$	for $10 < A \leq 100$ sq. ft
$(GC_p) = -0.8$	for $A > 100$ sq. ft

Figure 6-5B. Gabled and Hipped Roofs

Roofs $10° < \theta \leq 30°$

Positive with and without overhang: Zone 1,2,3

$(GC_p) = 0.5$	for $A \leq 10$ sq. ft
$(GC_p) = 0.7000 - 0.2000 \log A$	for $10 < A \leq 100$ sq. ft
$(GC_p) = 0.3$	for $A > 100$ sq. ft

Negative with and without overhang: Zone 1

$(GC_p) = -0.9$	for $A \leq 10$ sq. ft
$(GC_p) = -1.0000 + 0.1000 \log A$	for $10 < A \leq 100$ sq. ft
$(GC_p) = -0.8$	for $A > 100$ sq. ft

Negative without overhang: Zone 2 & 3

$(GC_p) = -2.1$	for $A \leq 10$ sq. ft
$(GC_p) = -2.8000 + 0.7000 \log A$	for $10 < A \leq 100$ sq. ft
$(GC_p) = -1.4$	for $A > 100$ sq. ft

Negative with overhang: Zone 2

$(GC_p) = -2.2$	for all A sq. ft

Negative with overhang: Zone 3

$(GC_p) = -3.7$	for $A \leq 10$ sq. ft
$(GC_p) = -4.9000 + 1.2000 \log A$	for $10 < A \leq 100$ sq. ft
$(GC_p) = -2.5$	for $A > 100$ sq. ft

Figure 6-5B. Gabled and Hipped Roofs

Roofs $30° < \theta \leq 45°$

Positive with and without overhang: Zone 1,2,3

$(GC_p) = 0.9$	for $A \leq 10$ sq. ft
$(GC_p) = 1.0000 - 0.1000 \log A$	for $10 < A \leq 100$ sq. ft
$(GC_p) = 0.8$	for $A > 100$ sq. ft

Negative with and without overhang: Zone 1

$(GC_p) = -1.0$	for $A \leq 10$ sq. ft
$(GC_p) = -1.2000 + 0.2000 \log A$	for $10 < A \leq 100$ sq. ft
$(GC_p) = -0.8$	for $A > 100$ sq. ft

Negative without overhang: Zone 2 & 3

$(GC_p) = -1.2$	for $A \leq 10$ sq. ft
$(GC_p) = -1.4000 + 0.2000 \log A$	for $10 < A \leq 100$ sq. ft
$(GC_p) = -1.0$	for $A > 100$ sq. ft

Negative with overhang: Zone 2 & 3

$(GC_p) = -2.0$	for $A \leq 10$ sq. ft
$(GC_p) = -2.2000 + 0.2000 \log A$	for $10 < A \leq 100$ sq. ft
$(GC_p) = -1.8$	for $A > 100$ sq. ft

Figure 6-6. Multispan Gabled Roofs with $h \le 60$ ft

Roofs $10° < \theta \le 30°$

Positive: Zone 1,2,3

$(GC_p) = 0.6$	for $A \le 10$ sq. ft
$(GC_p) = 0.8000 - 0.2000 \log A$	for $10 < A \le 100$ sq. ft
$(GC_p) = 0.4$	for $A > 100$ sq. ft

Negative: Zone 1

$(GC_p) = -1.6$	for $A \le 10$ sq. ft
$(GC_p) = -1.8000 + 0.2000 \log A$	for $10 < A \le 100$ sq. ft
$(GC_p) = -1.4$	for $A > 100$ sq. ft

Negative: Zone 2

$(GC_p) = -2.2$	for $A \le 10$ sq. ft
$(GC_p) = -2.7000 + 0.5000 \log A$	for $10 < A \le 100$ sq. ft
$(GC_p) = -1.7$	for $A > 100$ sq. ft

Negative: Zone 3

$(GC_p) = -2.7$	for $A \le 10$ sq. ft
$(GC_p) = -3.7000 + 1.0000 \log A$	for $10 < A \le 100$ sq. ft
$(GC_p) = -1.7$	for $A > 100$ sq. ft

Figure 6-6. Multispan Gabled Roofs with h ≤ 60 ft

Roofs 30° < θ ≤ 45°

Positive: Zone 1,2,3

$(GC_p) = 1.0$	for $A \leq 10$ sq. ft
$(GC_p) = 1.2000 - 0.2000 \log A$	for $10 < A \leq 100$ sq. ft
$(GC_p) = 0.8$	for $A > 100$ sq. ft

Negative: Zone 1

$(GC_p) = -2.0$	for $A \leq 10$ sq. ft
$(GC_p) = -2.9000 + 0.9000 \log A$	for $10 < A \leq 100$ sq. ft
$(GC_p) = -1.1$	for $A > 100$ sq. ft

Negative: Zone 2

$(GC_p) = -2.5$	for $A \leq 10$ sq. ft
$(GC_p) = -3.3000 + 0.8000 \log A$	for $10 < A \leq 100$ sq. ft
$(GC_p) = -1.7$	for $A > 100$ sq. ft

Negative: Zone 3

$(GC_p) = -2.6$	for $A \leq 10$ sq. ft
$(GC_p) = -3.5000 + 0.9000 \log A$	for $10 < A \leq 100$ sq. ft
$(GC_p) = -1.7$	for $A > 100$ sq. ft

Figure 6-7A. Monoslope Roofs with $h \leq 60$ ft

Roofs $3° < \theta \leq 10°$

Positive: All Zone

$(GC_p) = 0.3$	for $A \leq 10$ sq. ft
$(GC_p) = 0.4000 - 0.1000 \log A$	for $10 < A \leq 100$ sq. ft
$(GC_p) = 0.2$	for $A > 100$ sq. ft

Negative: Zone 1

$(GC_p) = -1.1$	for all A sq. ft

Negative: Zone 2

$(GC_p) = -1.3$	for $A \leq 10$ sq. ft
$(GC_p) = -1.4000 + 0.1000 \log A$	for $10 < A \leq 100$ sq. ft
$(GC_p) = -1.2$	for $A > 100$ sq. ft

Negative: Zone 2'

$(GC_p) = -1.6$	for $A \leq 10$ sq. ft
$(GC_p) = -1.7000 + 0.1000 \log A$	for $10 < A \leq 100$ sq. ft
$(GC_p) = -1.5$	for $A > 100$ sq. ft

Negative: Zone 3

$(GC_p) = -1.8$	for $A \leq 10$ sq. ft
$(GC_p) = -2.4000 + 0.6000 \log A$	for $10 < A \leq 100$ sq. ft
$(GC_p) = -1.2$	for $A > 100$ sq. ft

Negative: Zone 3'

$(GC_p) = -2.6$	for $A \leq 10$ sq. ft
$(GC_p) = -3.6000 + 1.0000 \log A$	for $10 < A \leq 100$ sq. ft
$(GC_p) = -1.6$	for $A > 100$ sq. ft

Figure 6-7A. Monoslope Roofs with $h \leq 60$ ft

Roofs $10° < \theta \leq 30°$

Positive: All Zone

$(GC_p) = 0.4$	for $A \leq 10$ sq. ft
$(GC_p) = 0.5000 - 0.1000 \log A$	for $10 < A \leq 100$ sq. ft
$(GC_p) = 0.3$	for $A > 100$ sq. ft

Negative: Zone 1

$(GC_p) = -1.3$	for $A \leq 10$ sq. ft
$(GC_p) = -1.5000 + 0.2000 \log A$	for $10 < A \leq 100$ sq. ft
$(GC_p) = -1.1$	for $A > 100$ sq. ft

Negative: Zone 2

$(GC_p) = -1.6$	for $A \leq 10$ sq. ft
$(GC_p) = -2.0000 + 0.4000 \log A$	for $10 < A \leq 100$ sq. ft
$(GC_p) = -1.2$	for $A > 100$ sq. ft

Negative: Zone 3

$(GC_p) = -2.9$	for $A \leq 10$ sq. ft
$(GC_p) = -3.8000 + 0.9000 \log A$	for $10 < A \leq 100$ sq. ft
$(GC_p) = -2.0$	for $A > 100$ sq. ft

Figure 6-7B. Sawtooth Roofs with $h \leq 60$ ft

Roofs $10° < \theta \leq 30°$

Positive: Zone 1

$(GC_p) = 0.7$ for $A \leq 10$ sq. ft
$(GC_p) = 0.8766 - 0.1766 \log A$ for $10 < A \leq 500$ sq. ft
$(GC_p) = 0.4$ for $A > 500$ sq. ft

Positive: Zone 2

$(GC_p) = 1.1$ for $A \leq 10$ sq. ft
$(GC_p) = 1.4000 - 0.3000 \log A$ for $10 < A \leq 100$ sq. ft
$(GC_p) = 0.8$ for $A > 100$ sq. ft

Positive: Zone 3

$(GC_p) = 0.8$ for $A \leq 10$ sq. ft
$(GC_p) = 0.9000 - 0.1000 \log A$ for $10 < A \leq 100$ sq. ft
$(GC_p) = 0.7$ for $A > 100$ sq. ft

Negative: Zone 1

$(GC_p) = -2.1$ for $A \leq 10$ sq. ft
$(GC_p) = -2.6886 + 0.5886 \log A$ for $10 < A \leq 500$ sq. ft
$(GC_p) = -1.1$ for $A > 500$ sq. ft

Negative: Zone 2

$(GC_p) = -3.2$ for $A \leq 10$ sq. ft
$(GC_p) = -4.1418 + 0.9418 \log A$ for $10 < A \leq 500$ sq. ft
$(GC_p) = -1.6$ for $A > 500$ sq. ft

Negative: Zone 3 (span A)

$(GC_p) = -4.1$ for $A \leq 10$ sq. ft
$(GC_p) = -4.5000 + 0.4000 \log A$ for $10 < A \leq 100$ sq. ft
$(GC_p) = -8.2782 + 2.2891 \log A$ for $100 < A \leq 500$ sq. ft
$(GC_p) = -2.1$ for $A > 500$ sq. ft

Negative: Zone 3 (spans BCD)

$(GC_p) = -2.6$ for $A \leq 100$ sq. ft
$(GC_p) = -4.6030 + 1.0015 \log A$ for $100 < A \leq 500$ sq. ft
$(GC_p) = -1.9$ for $A > 500$ sq. ft

Figure 6-8. Roof and Walls for Buildings with h > 60 ft

Roofs $\theta \leq 10°$

Negative: Zone 1

$(GC_p) = -1.4$	for $A \leq 10$ sq. ft
$(GC_p) = -1.6943 + 0.2943 \log A$	for $10 < A \leq 500$ sq. ft
$(GC_p) = -0.9$	for $A > 500$ sq. ft

Negative: Zone 2

$(GC_p) = -2.3$	for $A \leq 10$ sq. ft
$(GC_p) = -2.7120 + 0.4120 \log A$	for $10 < A \leq 500$ sq. ft
$(GC_p) = -1.6$	for $A > 500$ sq. ft

Negative: Zone 3

$(GC_p) = -3.2$	for $A \leq 10$ sq. ft
$(GC_p) = -3.7297 + 0.5297 \log A$	for $10 < A \leq 500$ sq. ft
$(GC_p) = -2.3$	for $A > 500$ sq. ft

Walls All θ

Positive: Zone 4 & 5

$(GC_p) = 0.9$	for $A \leq 20$ sq. ft
$(GC_p) = 1.1792 - 0.2146 \log A$	for $20 < A \leq 500$ sq. ft
$(GC_p) = 0.6$	for $A > 500$ sq. ft

Negative: Zone 4

$(GC_p) = -0.9$	for $A \leq 20$ sq. ft
$(GC_p) = -1.0861 + 0.1431 \log A$	for $20 < A \leq 500$ sq. ft
$(GC_p) = -0.7$	for $A > 500$ sq. ft

Negative: Zone 5

$(GC_p) = -1.8$	for $A \leq 20$ sq. ft
$(GC_p) = -2.5445 + 0.5723 \log A$	for $20 < A \leq 500$ sq. ft
$(GC_p) = -1.0$	for $A > 500$ sq. ft

2.4 ERRATA AND CLARIFICATION

Although the text, tables and figures of ASCE 7-95 were carefully reviewed prior to printing, certain errors inevitably went undetected. This section lists errors in the wind load provisions that have come to light as well as suggestions by users of the Standard to clarify certain provisions in the next edition.

- Page 1. At the bottom of the page, add the following footnote:

  * Asterisks appearing in Sections 1 through 8 and in Section 10 denote that additional information pertaining to the provision, figure, or table may be found in the Commentary.

  Also, for the same purpose, add an asterisk at the following locations: Section 6.1; Section 6.2 Building, enclosed; Building, open; Building, partially enclosed; Section 6.3; Figures 6-2, 6-3, 6-4, 6-5B, 6-5C, 6-7, 6-7b; and Tables 6-4, 6-5, 6-6, 6-7, 6-8, and 6-9.

- Fig. 6-2. Add Eq. 6-2 to this figure to make the procedure for wind speed-up over hills and escarpments more self-contained.

- Fig. 6-4. The title of this figure should read "...Main Wind-Force Resisting..."

- Fig. 6-4. Change Note 6 to read as follows: "Wind load calculations for the design of main wind-force resisting systems shall use a velocity pressure q_h based on Exposure C. For buildings sited within Exposure B, calculated pressures may be multiplied by 0.85."

- Fig. 6-5A, third footnote. Change 0° to 10°

- Fig. 6-5B. For $10° < \theta \leq 30°$, the uppermost solid line applies to zones 2 and 3 rather than to zones 1 and 2 as indicated.

- Fig. 6-7B. For the pressure coefficients portion of this figure, the intersection on the left y-axis of the negative coefficients GC_p is -2.2 for zone 1 and -3.2 for zone 2. Also, the last division on the x-axis is 1000 sq. ft (92.9 sq. m).

Chapter 3.
EXAMPLES

Several illustrative examples of the determination of wind loads using analytical procedures are developed to provide guidance to the reader. Specifically, six examples are presented for different configurations of buildings. The examples are as follows:

1. 100 ft x 200 ft x 160 ft high office building
2. Building as in Example 1 but placed on an escarpment
3. 200 ft x 250 ft x 20 ft eave height, gable roof (4:12) one story commercial/industrial building
4. Building as in Example 3 using alternative low-rise provisions for MWFRS
5. 40 ft x 80 ft x 15 ft high monoslope roof strip-mall building
6. 50 ft x 20 ft billboard sign placed 60 ft above ground

These examples represent a variety of situations in the determination of wind loads. The examples are worked out in detail to assist the reader in interpretation of the provisions of the Standard. Every effort has been made to check the accuracy of the numbers though no absolute assurance can be given.

3.1 EXAMPLE 1

In this example, design wind pressures for a typical office building are determined. The building data are as follows:

Location:	South Houston, Texas
Topography:	Flat
Terrain:	Suburban
Dimensions:	100 ft x 200 ft in plan
	Roof height of 157 ft with 3 ft parapet
	Flat roof
Framing:	Reinforced concrete rigid frame in both directions
	Floor and roof slabs provide diaphragm action
	Fundamental natural frequency is greater than 1 Hz
	(Since the height to least horizontal dimension is less than 4, the fundamental frequency is judged to be greater than 1 Hz.)
Cladding:	Mullions for glazing panels span 11 ft between floor slabs
	Mullion spacing is 5 ft
	Glazing panels are 5 ft wide x 5 ft 6 in. high (typical); they are resistant to wind-borne debris impact in the bottom 60 ft.

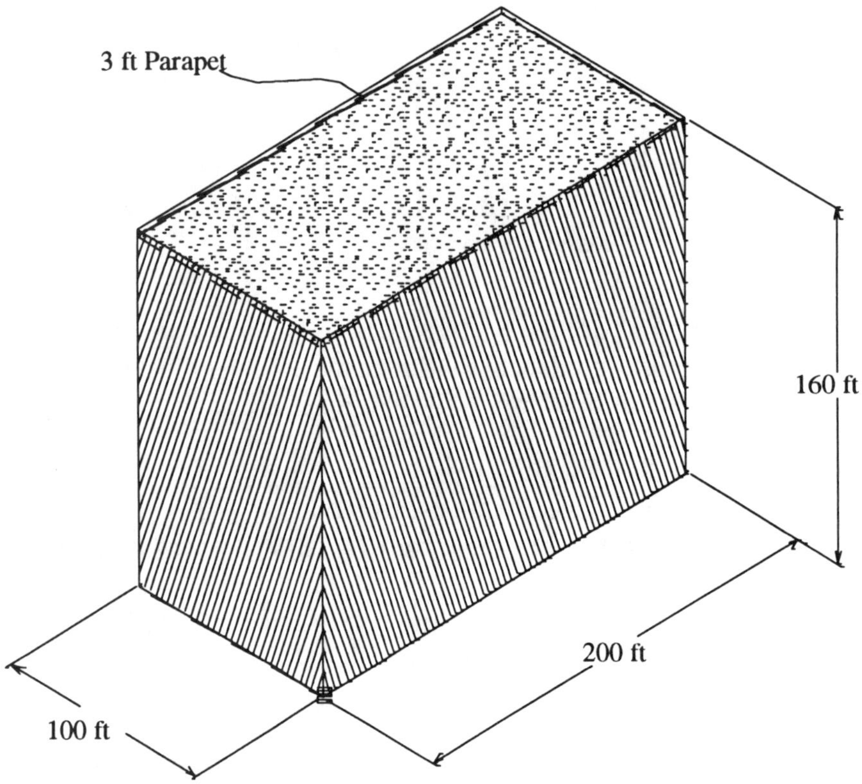

Figure 3.1 Building Dimensions for Example 1

Exposure and Building Classification

The building is located in a suburban area, therefore use Exposure B (Sec 6.5.3).

The building function is office space. It is not considered an essential facility or likely to be occupied by 300 persons in a single area at one time. Therefore, building category II is appropriate (Table 1-1).

Basic Wind Speed

Selection of the basic wind speed is addressed in Section 6.5.2 of the Standard. South Houston, Texas is located on the 120 mph contour. It is in a hurricane-prone region. The basic wind speed V = 120 mph (Fig. 6-1).

Velocity Pressures

The velocity pressures are computed using:

$$q_z = 0.00256 K_z K_{zt} V^2 I \text{ psf} \qquad \text{(Eq. 6-1)}$$

For this example, K_z is obtained from Table 6-3, K_{zt} = 1.0 (flat terrain), I = 1.0 for category II building (Table 6-2), and the basic wind speed is V = 120 mph. Substituting these values into the equation 6-1 yields:

$$q_z = 0.00256K_z(1.0)(120)^2(1.0)$$
$$q_z = 36.9\ K_z\ \text{psf}$$

Values for K_z and the resulting velocity pressures are given in the table below. The velocity pressure at mean roof height, q_h, is 41.3 psf.

K_z *Velocity pressures*

Height, ft	K_z	q_z, psf
0 - 15	0.57	21.0
30	0.70	25.8
50	0.81	29.9
80	0.93	34.3
120	1.04	38.3
Roof ht = 157	1.12	41.3
Parapet ht = 160	1.13	41.7

Design Wind Pressures for the Main Wind-Force Resisting System (MWFRS)

Equations for the design pressures for a building are given in Table 6-1 of the Standard. The equation for the MWFRS (the building is not flexible) is:

$$p = qGC_p - q_h(GC_{pi})$$

For the equation above
$q = q_z$ for windward wall at height z above ground
$q = q_h$ for leeward wall, side walls, and roof

Gust Effect Factor, G

The gust effect factor is given in Section 6.6.1.

For suburban terrain, Exposure B, use G = 0.80

Wall External Pressure Coefficients, C_p

The values for the external pressure coefficients for the various wall surfaces are obtained from Figure 6-3.

The windward wall pressure coefficient is 0.8.

The side wall pressure coefficient is - 0.7.

The leeward wall pressure coefficient is a function of the L/B ratio. For wind normal to the 200 ft face, L/B = 100/200 = 0.5, therefore the leeward wall pressure coefficient is - 0.5. For wind parallel to the 200 ft face, L/B = 200/100 = 2.0, therefore the leeward wall pressure coefficient is - 0.3.

Wall C_p

Surface	Wind Direction	L/B	C_p
Windward Wall	All	All	0.80
Leeward Wall	⊥ to 200 ft face	0.5	- 0.50
	ǁ to 200 ft face	2.0	- 0.30
Side Wall	All	All	- 0.70

Roof C_p (with the wind normal to the 200 ft face)

For h/L = 157/100 ≈ 1.6 > 1.0, and $\theta < 10^{\circ}$, two zones are specified:

0 to h/2, C_p = - 1.3

> h/2, C_p = - 0.7

The C_p = - 1.3 can be reduced with the area over which it is applicable.

Area = 200 x 79 = 15,800 sq ft

Reduction factor = 0.8

Reduced C_p = 0.8 x - 1.3 = - 1.04

Roof C_p

Distance from Leading Edge	C_p
0 to h/2	- 1.04
>h/2	- 0.70

* h = 157 ft

Roof C_p (with the wind parallel to the 200 ft face)

For h/L = 157/200 ≈ 0.8, interpolation in Figure 6-3 is required.

Roof C_p

Distance from Windward Edge	h/L ≤ 0.5	h/L = 0.8	h/L ≥ 1.0
0 to h/2	- 0.9	- 0.98	- 1.04
h/2 to h	- 0.9	- 0.78	- 0.7
h to 2h	- 0.5	- 0.62	- 0.7

Internal Pressure Coefficients ,(GC_{pi})

Values for GC_{pi} for buildings are addressed in Table 6-4 of the Standard.

It is assumed that the openings are evenly distributed in the walls. Although Houston, Texas is in a hurricane-prone region, wind-borne debris resistant glazing is used in the bottom 60 ft. Hence, from Table 6-4:

$$GC_{pi} = \pm 0.18$$

MWFRS Net Pressures

$$p = qGC_p - q_h(GC_{pi}) \qquad \text{(Table 6-1)}$$

Windward Wall Calculation at 30 ft, Wind Normal to 200 ft Face

$p = 25.8(0.80)(0.80) - 41.3(\pm 0.18)$

$p = 9.1$ psf with (+) internal pressure

$p = 23.9$ psf with (-) internal pressure

Roof Calculation for 0-79 ft from Edge, Wind Normal to 200 ft Face

$p = \ 41.3\ (0.8)\ (-1.04) - 41.3\ (\pm 0.18)$

$= -\ 41.8$ psf with (+) internal pressure

$= -\ 26.9$ psf with (-) internal pressure

Net Pressures for MWFRS: Wind Normal to 200 ft Face

Surface	z	q	C_p	Ext. Press.	Net Pressure (psf) With	
	ft	psf			($+GC_{pi}$)	($-GC_{pi}$)
Windward Wall	0-15	21.0	0.80	13.4	6.0	20.9
	30	25.8	0.80	16.5	9.1	24.0
	50	29.9	0.80	19.1	11.7	26.6
	80	34.3	0.80	21.9	14.5	29.4
	120	38.3	0.80	24.5	17.1	32.0
	157	41.3	0.80	26.4	19.0	33.9
Leeward Wall	All	41.3	-0.50	-16.5	-24.0	-9.1
Side Walls	All	41.3	-0.70	-23.1	-30.6	-15.7
Roof	0-79	41.3	-1.04	-34.4	-41.8	-26.9
	79-100	41.3	-0.70	-23.1	-30.6	-15.7

Notes: $q_h = 41.3$ psf; $G = 0.8$; $(GC_{pi}) = \pm 0.18$

Parapet load on MWFRS

For parapet load contribution to the MWFRS, the parapet may be approximated as a sign structure at ground level.

Force coefficient for $\nu = 3/200 < 3$, $C_f = 1.2$ (Table 6-8)

For open building and other structures the equation is give in Table 6-1

$$F = q_z \, GC_f \, A_f$$

If A_f is taken as 1 ft wide x 3 ft high,

$$F = 41.7\ (0.8)\ (1.2)\ (3) = 120 \text{ plf}$$

This force is to be applied on windward parapet and to the leeward parapet.

For design of parapet, see the loads on components and cladding.

Notes:

1. A similar loading with negative internal pressures may be considered; it will have reduced uplift on the roof and will not affect total horizontal shear
2. The load distribution steps on windward wall are the same as q_z

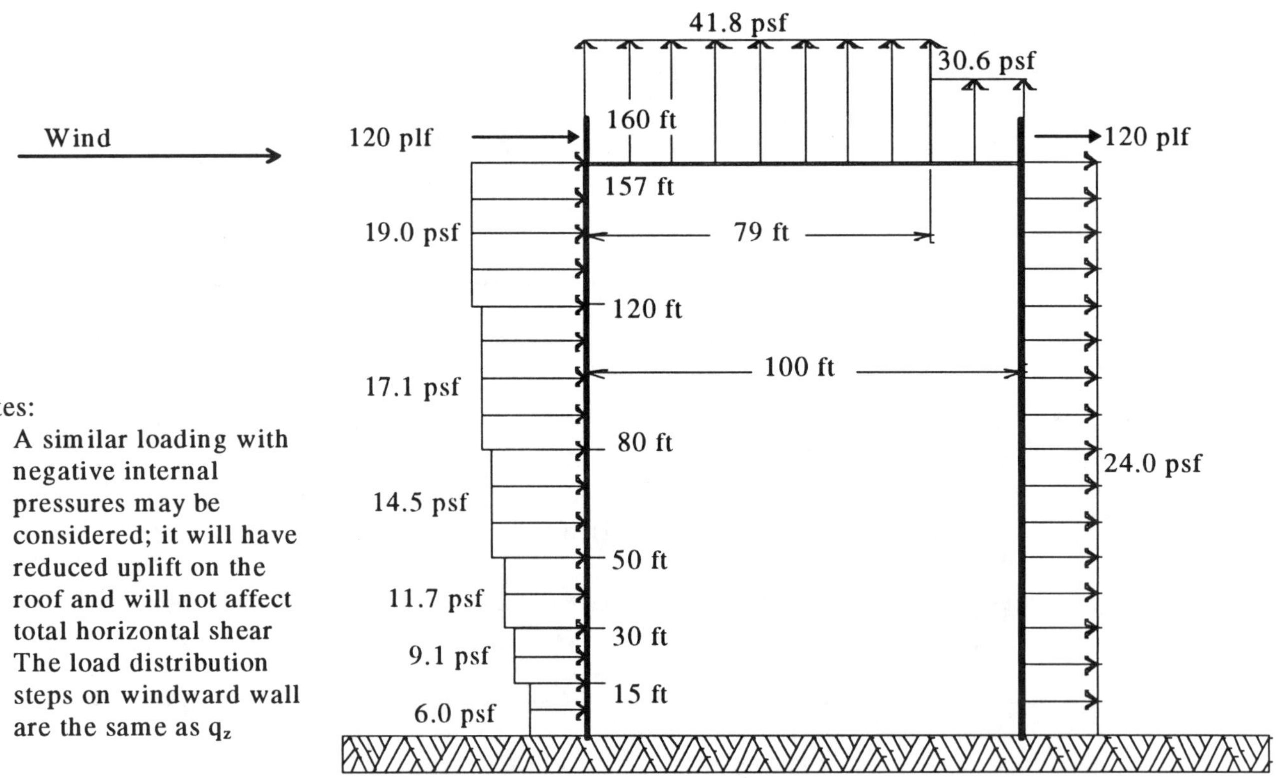

Figure 3.2 Net Pressures for MWFRS for Wind Normal to the 200 ft Face with Positive Internal Pressure

Net Pressures for MWFRS: Wind Parallel to 200 ft Face

Surface	z	q	C_p	Ext. Press.	Net Pressure (psf) With	
	ft	psf			$(+GC_{pi})$	$(-GC_{pi})$
Windward Wall	0-15	21.0	0.80	13.4	6.0	20.9
	30	25.8	0.80	16.5	9.1	24.0
	50	29.9	0.80	19.1	11.7	26.6
	80	34.3	0.80	21.9	14.5	29.4
	120	38.3	0.80	24.5	17.1	32.0
	157	41.3	0.80	26.4	19.0	33.9
Leeward Wall	All	41.3	- 0.30	- 9.9	- 17.4	-2.5
Side Walls	All	41.3	- 0.70	- 23.1	- 30.6	- 15.7
Roof	0-79	41.3	- 0.98	- 32.4	- 39.8	- 24.9
	79-157	41.3	- 0.78	- 25.8	- 33.2	-18.3
	157-200	41.3	- 0.62	- 20.5	- 27.9	- 13.0

Notes: q_h = 41.3 psf; G = 0.8; $(GC_{pi}) = \pm 0.18$

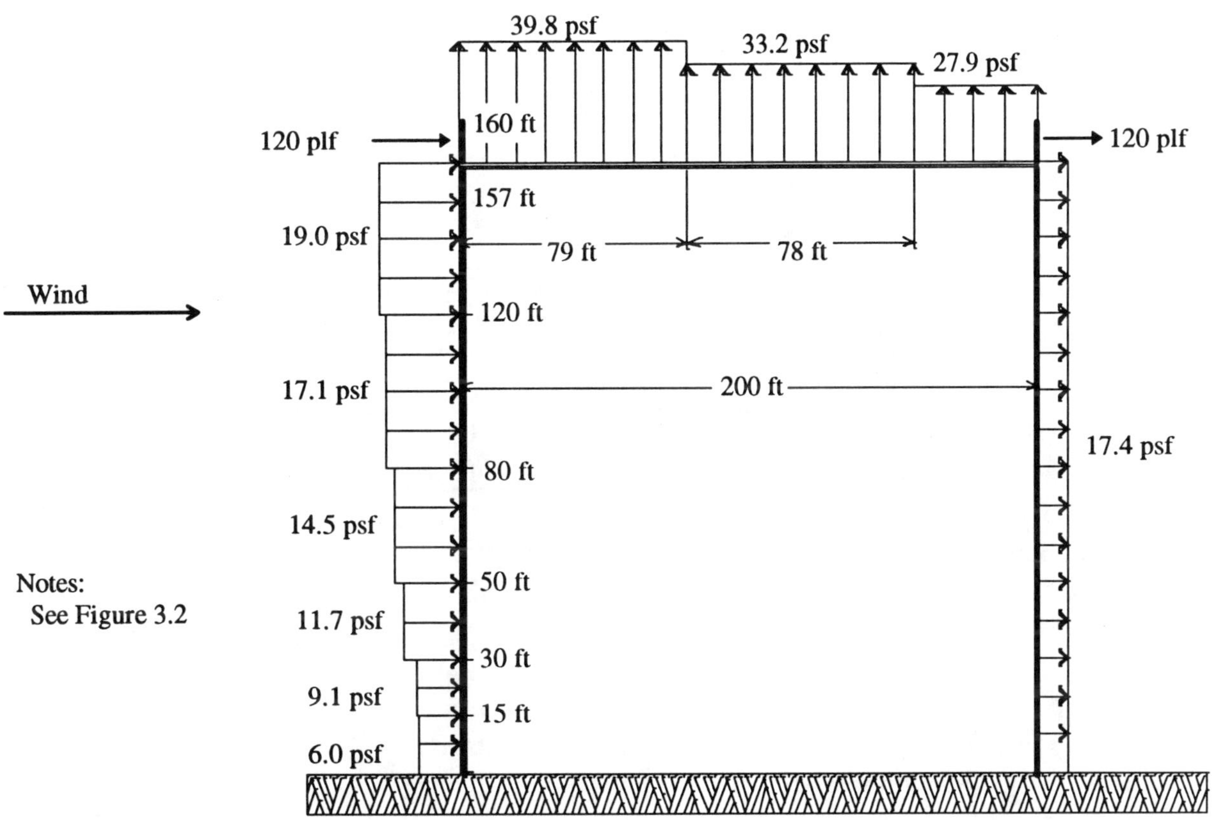

Figure 3.3 Net Pressures for MWFRS for Wind Parallel to the 200 ft Face with Positive Internal Pressure

Design Pressures for Components and Cladding

Design pressure equation for components and cladding is given in Table 6-1 of the Standard. The equation is:

$$p = q[(GC_p) - (GC_{pi})]$$

where:
$q = q_z$ for positive pressures at height z
$q = q_h$ for negative pressures
(GC_p) = external pressure coefficient (Figure 6-8)

Wall Design Pressures

The pressure coefficients (GC_p) are a function of effective wind area. The definition of effective wind area for a component or cladding panel is the span length multiplied by an effective width that need not be less than one third the span length (Section 6.2). The effective wind areas, A for wall components are:

Mullion: A = 11(5) = 55 sq. ft (controls)
or = 11(11/3) = 40.3 sq. ft

Glazing Panel: A = 5(5.5) = 27.5 sq. ft (controls)
or = 5(5/3) = 8.3 sq. ft

Width of Zone 5: a = 0.1(100) = 10 ft.

Wall (GC_p)

Component	A (sq. ft)	GC_p zone 4&5 $(+GC_p)$	GC_p zone 4 $(-GC_p)$	GC_p zone 5 $(-GC_p)$
Mullion	55	0.81	- 0.84	- 1.55
Panel	27.5	0.87	- 0.88	- 1.72

The internal pressure coefficient, $(GC_{pi}) = \pm 0.18$.

Typical Design Pressure Calculations

Mullion in zone 4 (Negative Pressure)

$p = 41.3[(-0.84)-(\pm 0.18)]$
= - 42.1 with positive internal pressure (controls)
= - 27.3 with negative internal pressure

Controlling negative pressure is obtained with positive internal pressure and controlling positive pressure is obtained with negative internal pressure.

Mullion Design Pressures, psf

Component	z	Design Pressures			
		zone 4		zone 5	
	ft	Positive	Negative	Positive	Negative
Mullion	0-15	20.8	- 42.1	20.8	- 71.4
	15-30	25.5	- 42.1	25.5	- 71.4
	30-50	29.6	- 42.1	29.6	- 71.4
	50-80	33.9	- 42.1	33.9	- 71.4
	80-120	38.0	- 42.1	38.0	- 71.4
	120-157	40.9	- 42.1	40.9	- 71.4

Panel Design Pressures, psf

Component	z	Design Pressures			
		zone 4		zone 5	
	ft	Positive	Negative	Positive	Negative
Panel	0-15	22.1	- 43.8	22.1	- 78.4
	15-30	27.1	- 43.8	27.1	- 78.4
	30-50	31.4	- 43.8	31.4	- 78.4
	50-80	36.0	- 43.8	36.0	- 78.4
	80-120	40.2	- 43.8	40.2	- 78.4
	120-157	43.4	- 43.8	43.4	- 78.4

Parapet Design Pressures

Although the Standard does not provide specific design pressures for parapets, the values of GC_p given in Figure 6-8 allow for rational assessment of these pressures. For a parapet on the windward building face the net pressure is assumed to be algebraic sum of the positive pressure for walls (zone 4 & 5) and the negative pressure for roof zone 2. For a parapet on the leeward or corner of the building face, the net pressure is assumed to be the algebraic sum of the positive pressure for walls (zones 4 & 5) and the negative pressure for walls (zone 4 & 5). The effective wind area will depend upon the dimensions and framing details of the parapet. In this example, the effective wind area is assumed to be 3 ft x 3 ft = 9 sq ft.

Windward Parapet:

$$p = 41.7(0.9) - 41.3(-2.3)$$
$$= 132.5 \text{ psf (Directed inward)}$$

Leeward or Corner Area Parapet:

Zone 4:

$$p = 41.7(0.9) - 41.3(-0.9)$$
$$= 74.7 \text{ psf (Directed outward)}$$

Zone 5:

$$p = 41.7(0.9) - 41.3(-1.8)$$
$$= 111.9 \text{ psf (Directed outward)}$$

Roof Design Pressures

The component and cladding roof pressure coefficients are given in Figure 6-8. The pressure coefficients are a function of the effective wind area. Since, specific components of roof are not identified, design pressures are given for various effective wind areas, A.

Roof (GC_p)

	External Pressure Coefficient	
A sq. ft	zone 1 ($-GC_p$)	zone 2 and 3 ($-GC_p$)*
≤ 10	-1.40	-2.30
20	-1.31	-2.18
100	-1.11	-1.89
250	-0.99	-1.72
400	-0.93	-1.64
≥ 500	-0.90	-1.60

* Note 7 in Figure 6-8 permits treatment of zone 3 as zone 2 if parapet of 3 ft or higher is provided.

The design pressures are the algebraic sum of external and internal pressures. Positive internal pressure provide controlling negative pressures. These design pressures act across the roof surface (interior to exterior).

Roof Design Pressures, psf

	Design Pressures Negative	
A sq. ft	zone 1	zone 2 and 3
≤ 10	-65.2	-102.4
20	-61.5	-97.4
100	-53.3	-85.5
250	-48.3	-78.4
400	-45.8	-75.1
500	-44.6	-73.5

3.2 EXAMPLE 2

In this example, velocity pressures for the office building of **Example 1**, when it is located on an escarpment, are determined. Design pressures for MWFRS and components and cladding can be determined in the same manner as **Example 1** once velocity pressures, q_z and q_h are determined.

Location:	City in Alaska
Topography:	Escarpment as shown
Terrain:	Suburban
Dimensions:	100 ft x 200 ft in plan
	Roof height of 157 ft with 3 ft parapet
	Flat roof
Framing:	Reinforced concrete rigid frame in both directions
	Floor and roof slabs provide diaphragm action
	Fundamental natural frequency is greater than 1 Hz
Cladding:	Mullions for glazing panels span 11 ft between floor slabs
	Mullion spacing is 5 ft
	Glazing panels are 5 ft wide x 5 ft 6 in. high (typical). Glazing does not have to be wind-borne debris impact resistant because Alaska is not in a hurricane-prone region (note 5 of Table 6-4).

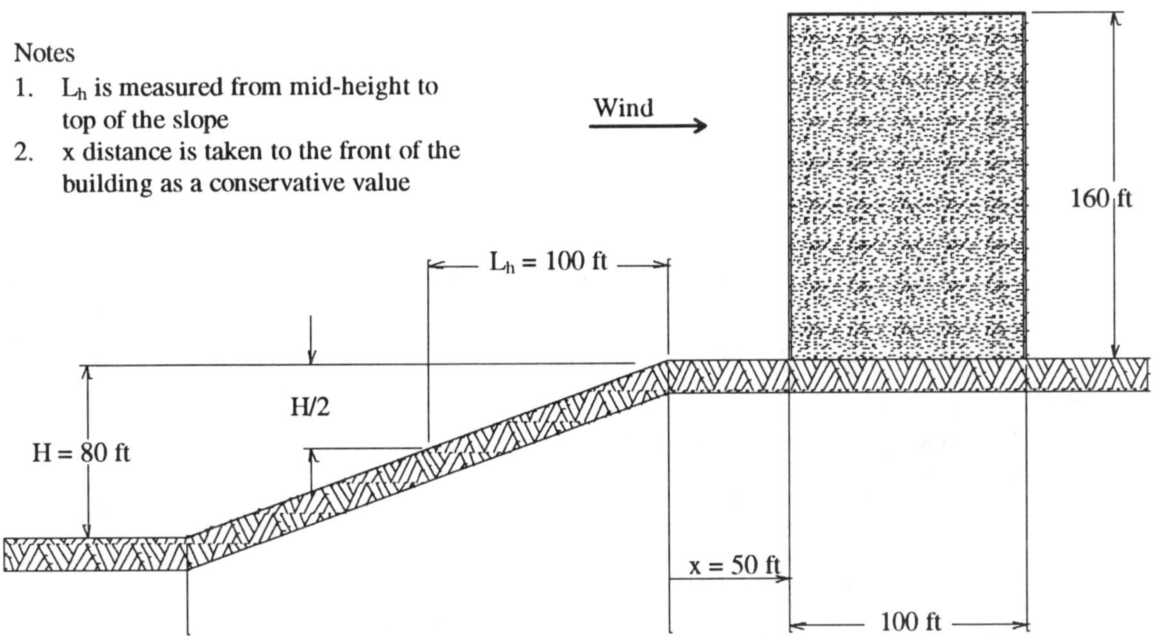

Figure 3.4 Office Building on Escarpment

Exposure, Building Classification and Basic Wind Speed

Same as Example 1: Exposure B
Category II
V = 120 mph (same as Example 1)

Velocity Pressures

The velocity pressure equation is:

$$q_z = 0.00256 K_z K_{zt} V^2 I \text{ psf}$$

For this example, K_z is obtained from Table 6-3 and K_{zt} is determined using Figure 6-2. I = 1.0 for Category II and V = 120 mph.

Determination of K_{zt}

The topographic effect of escarpment applies only when the upwind terrain is free of topographic features for a distance equal to 50H or 1 mile, whichever is smaller. For this example, it is assumed that there are no topographic features upwind for a distance of 4000 ft.

For use in Figure 6-2 of the Standard:

H = 80 ft
L_h = 100 ft
x = 50 ft (distance to the front face of the building)

Since $\frac{H}{L_h} = 0.8 > 0.5$

Note 2: use $\frac{H}{L_h} = 0.5$ and $L_h = 2H = 160$ ft

The building is on 2-D escarpment.

For $\frac{H}{L_h} = 0.5$; $K_1 = 0.43$ (Figure 6-2)

For $\frac{x}{L_h} = 0.31$; $K_2 = 0.92$ (Figure 6-2)

(Note: In Figure 6-2 *All Other Cases* for K_z refers to Ridge and Hill)

K_3 depends on $\frac{z}{L_h}$ (tabulated below)

$K_{zt} = (1+K_1K_2K_3)^2$ (equation 6-2 of the Standard)

$q_z = 0.00256 K_z K_{zt} (120)^2 (1.0)$

Velocity Pressures, psf

Height, ft	K_z	z/L_h*	K_3	K_{zt}	q_z, psf
0-15	0.57	0.05	0.89	1.82	38.3
30	0.70	0.14	0.71	1.64	42.4
50	0.81	0.25	0.54	1.47	43.9
80	0.93	0.41	0.36	1.31	45.0
120	1.04	0.63	0.20	1.16	44.5
h = 157	K_h = 1.12	0.87	0.12	1.10	45.5

* z is taken mid-way between the height range because it is unconservative to take top height of the range. Note; L_h = 160 ft

Effect of Escarpment

Velocity pressures q_z are compared with the values of Example 1 to assess the effect of the escarpment. The increase in velocity pressures does not directly translate into an increase in design pressures as discussed below.

Velocity Pressure q_z, psf

Height, ft	Homogeneous Terrain Example 1	Escarpment Example 2	% increase
0-15	21.0	38.3	82
30	25.8	42.4	64
50	29.9	43.9	47
80	34.3	45.0	31
120	38.3	44.5	16
157 (roof)	41.3	45.5	10

For MWFRS, the windward wall pressures will increase by the percentages shown at various heights; however the leeward wall, side wall, roof and internal pressures will increase by 10% since these pressures are controlled by velocity pressure at roof height, q_h.

For components and cladding, the negative (outward acting) design pressures will also increase only by 10%.

3.3 EXAMPLE 3

In this example, design wind pressures for a large one-story commercial-industrial building are determined. The building data are as follows:

Location:	Memphis, Tennessee
Terrain:	Flat farmland
Dimensions:	200 ft x 250 ft in plan
	Eave height of 20 ft
	Roof Slope 4:12 (18.4 degrees)
Framing:	Rigid frame spans the 200 ft direction
	Rigid frame spacing is 25 ft
	Cross bracing in 250 ft direction
	Girts and purlins span between rigid frames (25 ft span)
	Girt spacing is 6 ft 8 in
	Purlin Spacing is 5 ft
Cladding:	Roof panel dimensions are 2 ft x 20 ft
	Roof fastener spacing is 1 ft on center
	Wall panel dimensions are 2 ft x 20 ft
	Wall fastener spacing is 1 ft on center
	Openings uniformly distributed

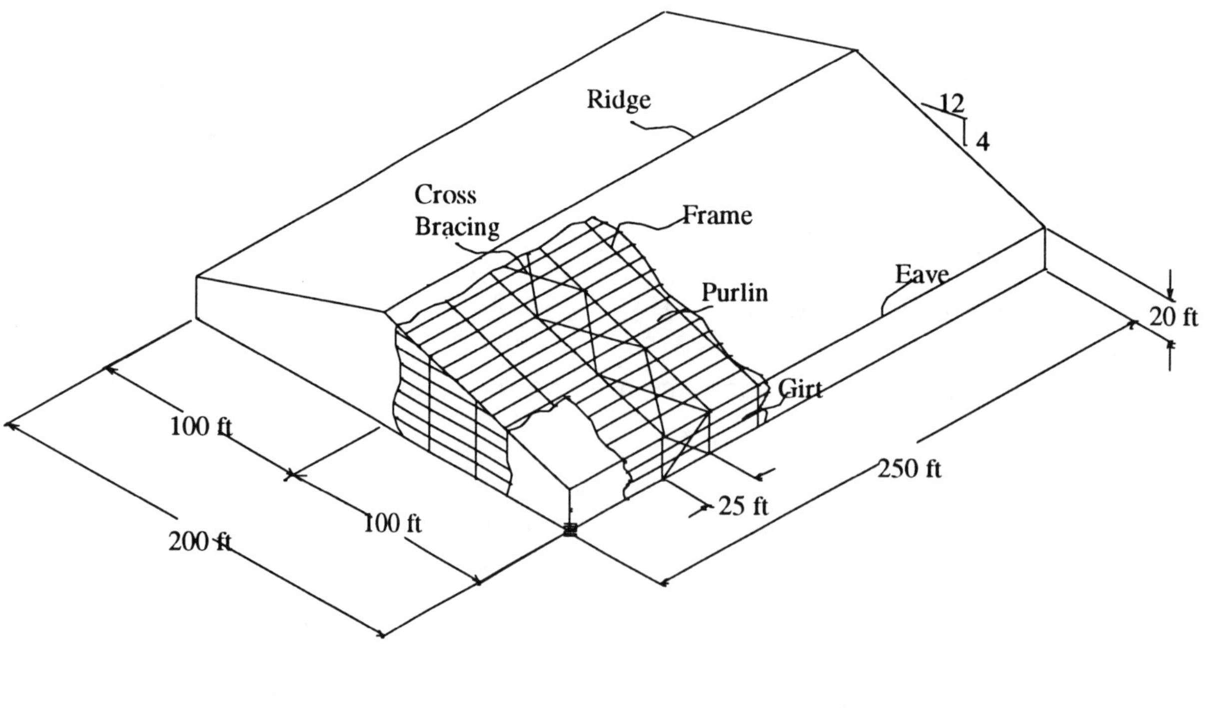

Figure 3.5 Dimensions and Framing of the building of Example 3

Exposure and Building Classification

The building is located in flat open farmland; therefore exposure category C.

The Building function is industrial-commercial. It is not considered an essential facility or likely to be occupied by 300 persons at one time. Category II is appropriate (Table 1-1).

Basic Wind Speed

Selection of basic wind speed is addressed in Section 6.5.2 of the Standard. Memphis, Tennessee is neither located in the special wind region nor is there any reason to suggest that winds at the site are unusual and require additional attention. Therefore, the basic wind speed is V = 90 mph (Fig. 6-1).

Velocity Pressures

The velocity pressures are computed using:

$$q_z = 0.00256 K_z K_{zt} V^2 I \text{ psf} \qquad \text{(Eq. 6-1)}$$

For this example, K_z is obtained from Table 6-3, K_{zt} = 1.0 (no topographic effect), I = 1.0 for category II building, and the basic wind speed is V = 90 mph. Substituting these values into the equation 6.1 yields:

$$q_z = 0.00256 K_z (1.0)(90)^2 (1.0)$$
$$q_z = 20.7 K_z \text{ psf}$$

Values for K_z and the resulting velocity pressures are as shown.
Mean roof height h = 36.7 ft.

Velocity Pressures, psf

Height, ft	K_z	q_z, psf
0 - 15	0.85	17.6
Eave ht = 20	0.90	18.7
30	0.98	20.3
h = 36.7	1.02	21.2*
40	1.04	21.6
50	1.09	22.6
Ridge ht = 53.3	1.10	22.8

* q_h = 21.2 psf

Design Wind Pressures for the Main Wind-Force Resisting System (MWFRS)

Equations for the design pressures and forces for a building or other structures are given in Table 6-1 of the Standard. The equation for MWFRS for buildings of all heights is:

$$p = qGC_p - q_h(GC_{pi})$$

where
$q = q_z$ for windward wall at height z above the ground
$q = q_h$ for leeward wall, side walls, and roof
G = 0.85 for Exposure C (Section 6.6.1)
C_p = values obtained from Figure 6-3
(GC_{pi}) = values obtained from Table 6-4

There are two alternative procedures to determine wind pressures for MWFRS of low-rise buildings. Pressure coefficients shown in Figure 6-3 are applicable for buildings of all heights; while pressure coefficients given in Figure 6-4 are applicable to low-rise buildings with mean roof height, h less or equal to 60 ft. Pressure coefficients of Figure 6-3 are used here. The procedure to use Figure 6-4 is illustrated in Example 4.

For this example, when the wind is normal to the ridge, the windward roof experiences both positive and negative external pressures. Combining these external pressures with positive and negative internal pressures will result in four loading cases when wind is normal to the ridge.

When wind is parallel to the ridge, positive and negative internal pressures result in two loading cases.

Wall C_p from Figure 6-3

The pressure coefficients for the windward wall and for the side walls are 0.8 and - 0.7, respectively, for all L/B ratios.

The leeward wall pressure coefficient is a function of the L/B ratio. For wind normal to the ridge, L/B = 200/250 = 0.8; therefore the leeward wall pressure coefficient is - 0.5. For flow parallel to the ridge, L/B = 250/200 = 1.25; the value of C_p is linearly interpolated. The wall pressure coefficients are summarized below.

Wall C_p

Surface	Wind Direction	L/B	C_p
Windward Wall	All	All	0.80
Leeward Wall	⊥ to ridge	0.8	- 0.50
	∥ to ridge	1.25	- 0.45*
Side Wall	All	All	- 0.70

* By linear interpolation

Roof C_p from Figure 6-3 (Wind Normal to Ridge)

The roof pressure coefficients for the MWFRS are obtained from the lower table in Figure 6-3. For the roof angle of 18.4 degrees, linear interpolation in this table is used to establish C_p. For wind normal to the ridge, h/L = 36.7/200 = 0.18.

Roof C_p (Wind Normal to Ridge)

Surface	15^0	18.4^0	20^0
Windward Wall	- 0.5	- 0.36*	- 0.3
	0.0	0.14*	0.2
Leeward Wall	- 0.5	- 0.57*	- 0.6

*by linear interpolation

Internal GC_{pi}

Values for GC_{pi} for buildings are addressed in Table 6-4 of the Standard.

The openings are evenly distributed in the walls and Memphis, TN is not in a hurricane-prone region; therefore

$$GC_{pi} = \pm 0.18$$

MWFRS Net Pressures

$p = qGC_p - q_h(GC_{pi})$
$p = q(0.85)C_p - 21.2\ (\pm 0.18)$
where
$q = q_z$ for windward wall
$q = q_h$ for leeward wall, side wall, and roof

Typical Calculation

Windward Wall, 0-15 ft, Wind Normal to Ridge

$p = 17.6(0.85)(0.8) - 21.1(\pm 0.18)$
p = 8.2 psf with (+) internal pressure
p = 15.8 psf with (-) internal pressure

The net pressures for the MWFRS are summarized in the following tables.

MWFRS Pressures: Wind Normal to Ridge

Surface	z ft	q, psf	C_p	Net Pressure psf with (+GCpi)	Net Pressure psf with (-GCpi)
Windward Wall	0-15	17.6	0.8	8.2	15.8
	20	18.7	0.8	8.9	16.5
Leeward Wall	All	21.2	- 0.5	- 12.8	- 5.2
Side Walls	All	21.2	- 0.7	- 16.4	- 8.8
Windward Roof*	-	21.2	- 0.36	- 10.3	- 2.7
			0.14	- 1.3	6.3
Leeward Roof	-	21.2	- 0.57	- 14.1	- 6.4

Notes: q_h = 21.2 psf; G = 0.085; (GC_{pi}) = ± 0.18

*Two loadings on windward roof and two internal pressures give a total of four loading cases (see Figures 3.6 and 3.7)

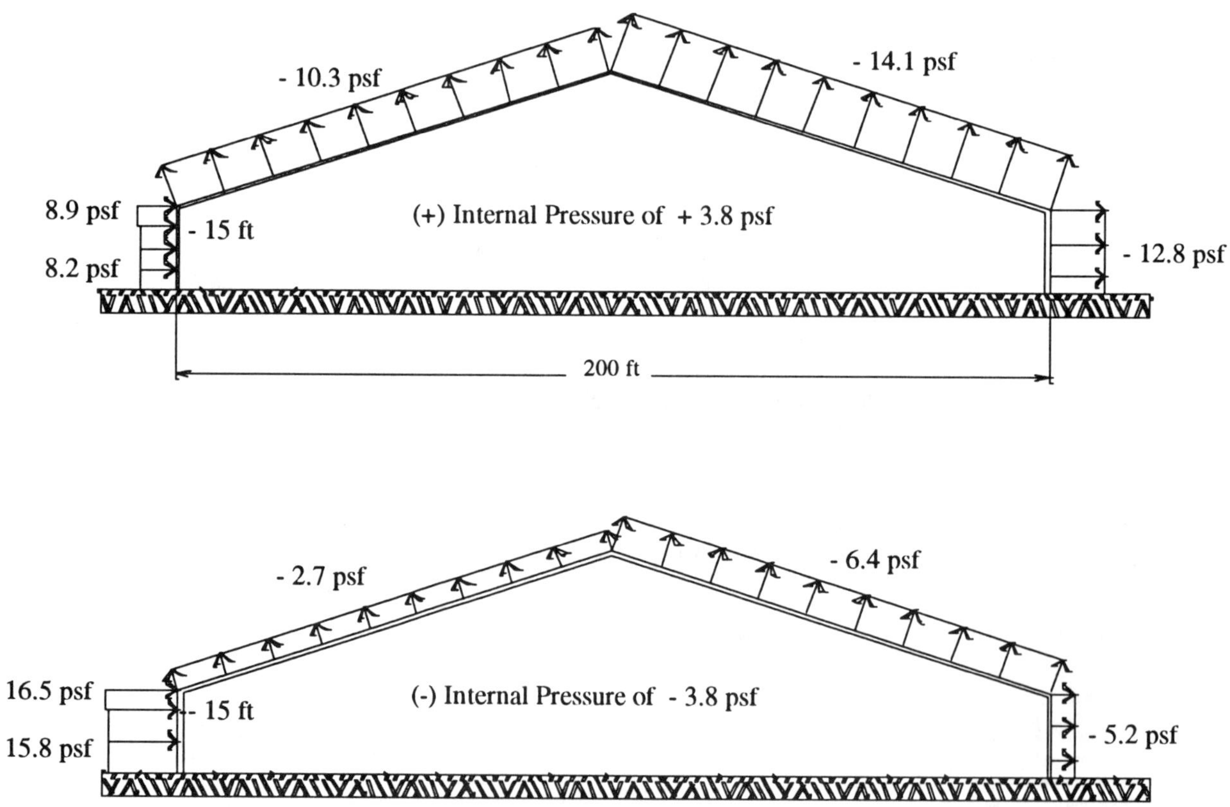

Figure 3.6 Net Design Wind Pressures for MWFRS when Wind is Normal to Ridge with Negative Windward External Roof Pressure Coefficient

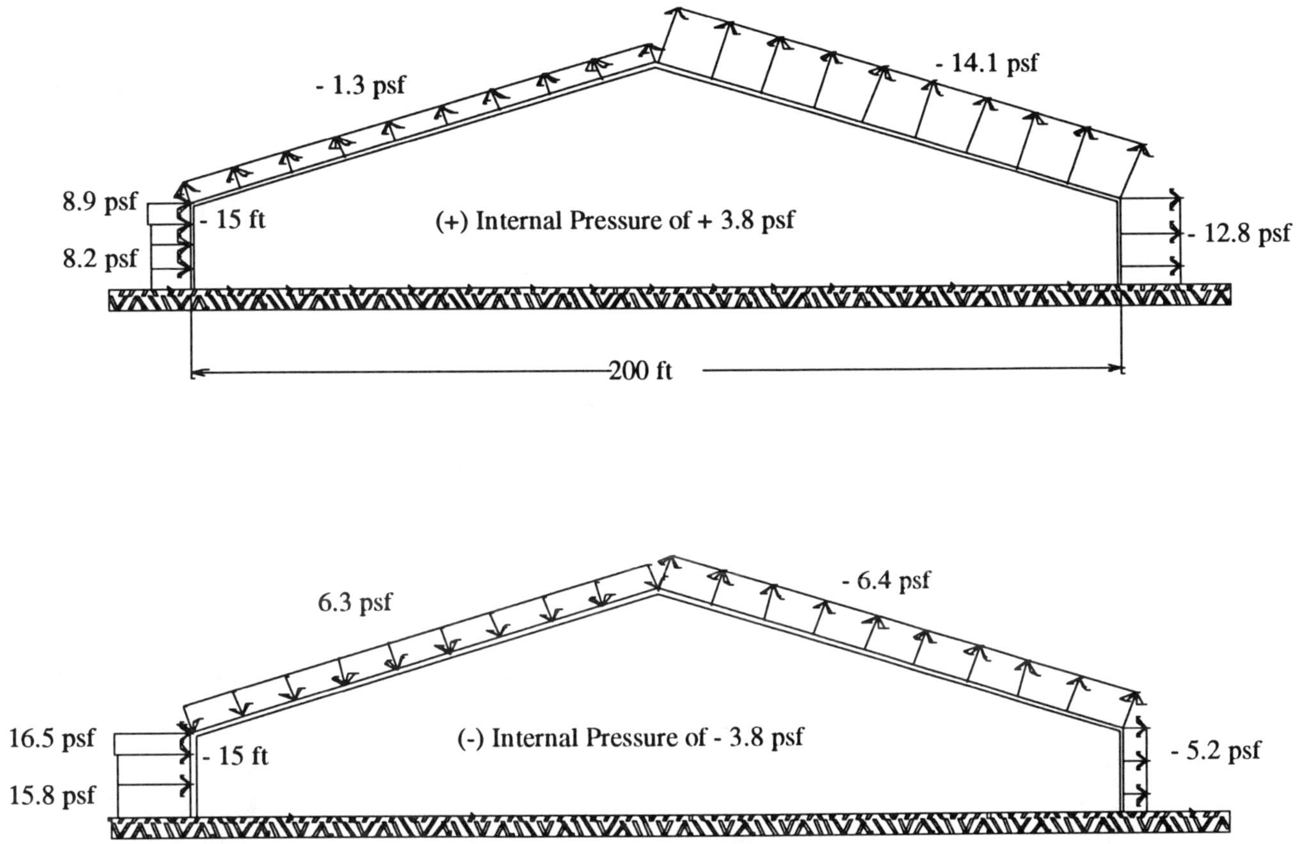

Figure 3.7 Net Design Wind Pressures for MWFRS when Wind is Normal to Ridge with Positive Windward External Roof Pressure Coefficient

Roof C_p from Figure 6-3 (Wind Parallel to Ridge)

For wind parallel to the ridge, h/L = 36.7/250 = 0.147. The values of C_p for wind parallel to ridge for all θ are obtained from Fig. 6-3 and are shown below.

Roof C_p (Wind Parallel to Ridge)

Surface	h/L	Distance from Windward Edge	C_p
Roof	≤ 0.5	0 to h	- 0.9
		h to 2h	- 0.5
		> 2h	- 0.3

MWFRS Pressures: Wind Parallel to Ridge

Surface	z ft	q psf	C_p	Net Pressure psf with (+GC_{pi})	(-GC_{pi})
Windward Wall	0-15	17.6	0.8	8.2	15.8
	20	18.7	0.8	8.9	16.5
	30	20.3	0.8	10.0	17.6
	40	21.6	0.8	10.9	18.5
	53.3	22.8	0.8	11.7	19.3
Leeward Wall	All	21.2	- 0.45	- 11.9	- 4.3
Side Walls	All	21.2	- 0.7	- 16.4	- 8.8
Roof	0 to h*	21.2	- 0.9	- 20.0	- 12.4
	h to 2h*	21.2	- 0.5	- 12.8	- 5.2
	> 2h*	21.2	- 0.3	- 9.2	- 1.6

Notes: q_h = 21.2 psf; G = 0.85; (GC_{pi}) = ± 0.18; h = 36.7
* Distance from windward edge

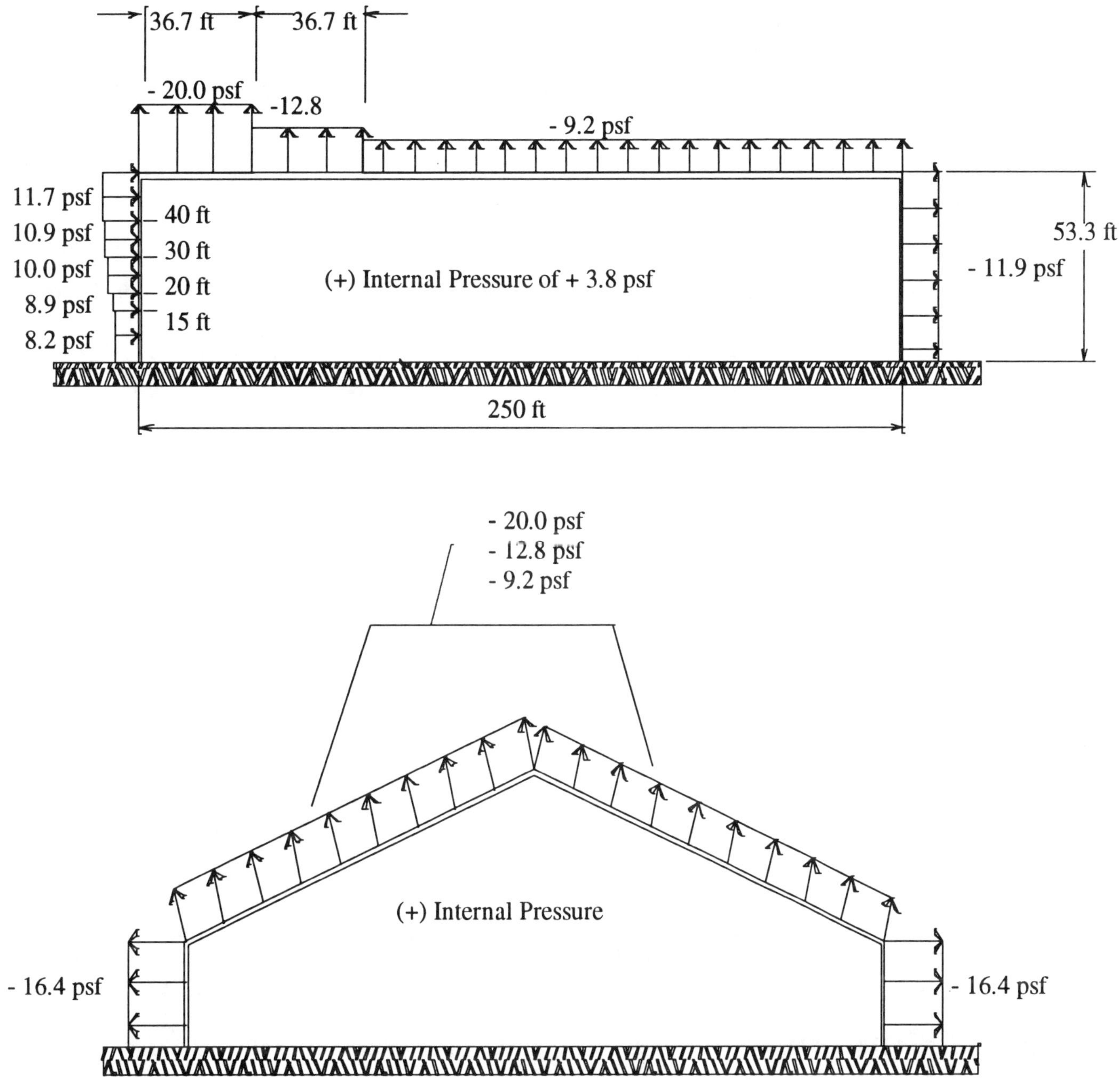

(a) With Positive Internal Pressure

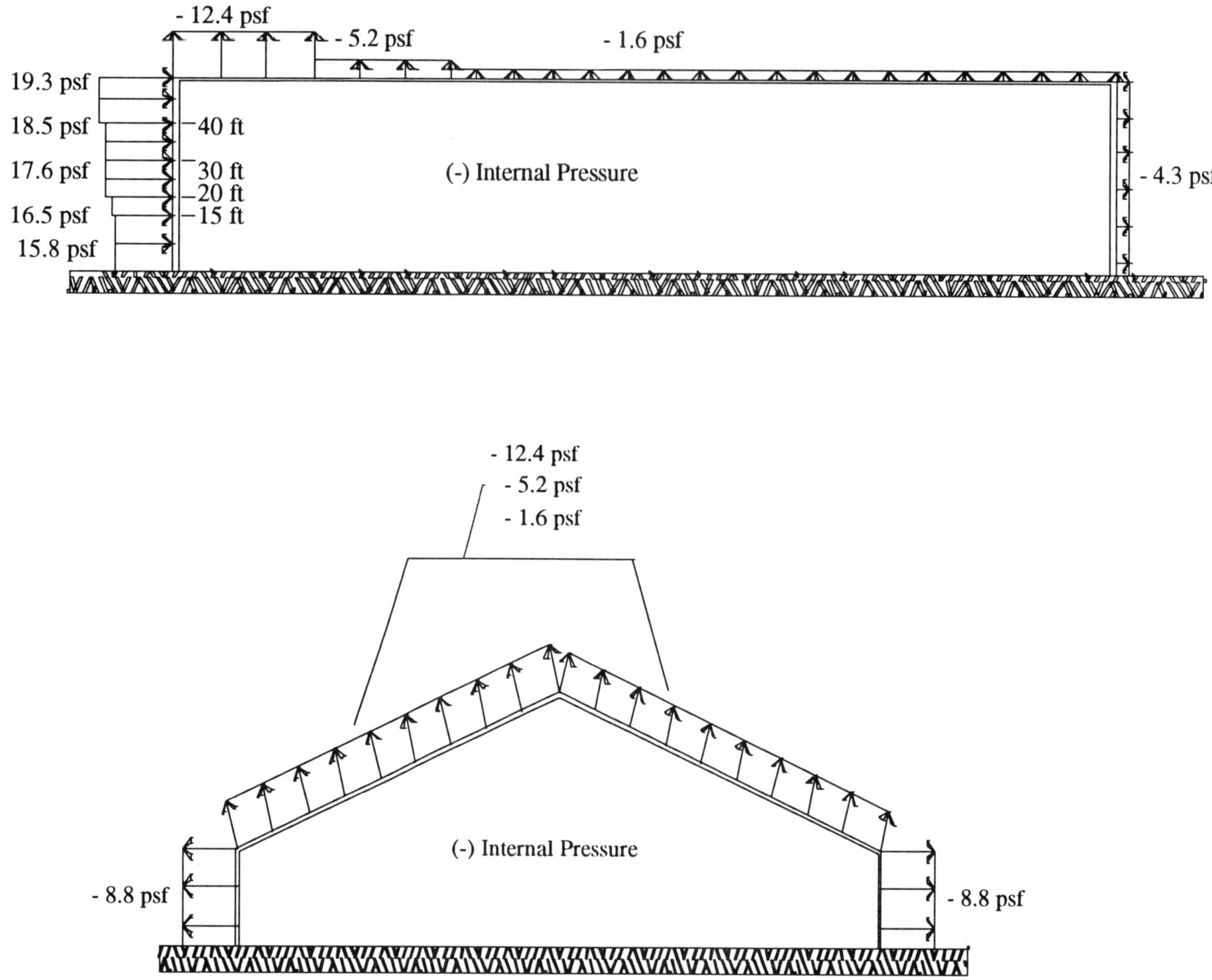

(b) With Negative Internal Pressure

Figure 3.8 Net Design Wind Pressures for MWFRS when Wind is Parallel to Ridge

Design Pressures for Components and Cladding (C&C)

Equations for the design pressures for components and cladding are given in Table 6-1 of the Standard. The equation for buildings with $h \leq 60$ ft is:

$$p = q_h[(GC_p) - (GC_{pi})]$$

where:

$q_h = 21.2$ psf

(GC_p) = values obtained from Figure 6-5

$(GC_{pi}) = \pm 0.18$ for this building

Wall Components and Cladding Pressures

The pressure coefficients (GC_p) are a function of effective wind area. The definitions of effective wind area for a component or cladding panel is the span length multiplied by an effective width that need not be less than one third the span length; however, for a fastener it is the area tributary to an individual fastener.

Girt: A = 25(6.67) = 167 sq. ft
or = 25(25/3) = 208 sq. ft (controls)

Panel: A = 6.67(2) = 13.3 sq. ft
or = 6.67(6.67/3) = 14.8 sq. ft (controls)

Fastener: A = 6.67(1) = 6.7 sq. ft

Wall Coefficients (GC_p) from Figure 6-5A

C&C	A sq. ft	External (GC_p) Zones 4&5	Zone 4	Zone 5
Girt	208	0.77	- 0.87	- 0.93
Panel	14.8	0.97	- 1.07	- 1.34
Fastener	6.7	1.00	- 1.10	- 1.40
Other	≤ 10	1.00	- 1.10	- 1.40
Other	≥ 500	0.70	- 0.80	- 0.80

*Other C&C can be doors, windows, etc.

Typical calculations of design pressures for Girt in Zone 4

For maximum negative pressure
p = 21.2[(-0.87) - (± 0.18)]
p = -22.2 psf with positive internal pressure (controls)
p = -14.6 psf with negative internal pressure

For maximum positive pressure
p = 21.2[(0.77) - (± 0.18)]
p = 12.5 psf with positive internal pressure
p = 20.1 psf with negative internal pressure (controls)

Net Wall Component Pressures, psf

Component	Controlling Design Pressures, psf			
	Zone 4		Zone 5	
	positive	negative	positive	negative
Girt	20.1	- 22.1	20.1	- 23.5
Panel	24.4	- 26.5	24.4	- 32.2
Fastener	25.0	- 27.1	25.0	- 33.5
A ≤ 10 sq ft	25.0	- 27.1	25.0	- 33.5
A ≥ 500 sq ft	18.7	- 20.8	18.7	- 20.8

Roof Components and Cladding Pressures

Effective wind areas of roof C&C

Purlin: A = 25(5) = 125 sq. ft
or = 25(25/3) = 208 sq. ft (controls)

Panel: A = 5(2) = 10 sq. ft (controls)
or = 5(5/3) = 8.3 sq. ft

Fastener: A = 5(1) = 5 sq. ft

Roof Coefficients, (GC_p) from Figure 6-5B; $10 < \theta \leq 30$ degrees

Component	A sq. ft	External (GC_p) Zones 1,2,3	Zone 1	Zones 2&3
Purlin	208	0.3	-0.8	- 1.4
Panel	10	0.5	-0.9	- 2.1
Fastener	5	0.5	-0.9	- 2.1
Other*	≤ 10	0.5	-0.9	- 2.1
Other	≥ 100	0.3	-0.8	- 1.4

* Other C&C can be skylite, etc.

Typical calculations of design pressures for Purlin in zone 1:

For maximum negative pressure

p = 21.2[(-0.8) - (± 0.18)]

p = -20.7 with positive internal pressure (controls)

p = -13.1 with negative internal pressure

For maximum positive pressure

p = 21.2[(0.3) - (± 0.18)]

p = 2.5 with positive internal pressure

p = 10.2 with negative internal pressure (controls)

Net Roof Component Pressures, psf

	Controlling Design Pressures, psf		
Component	Positive	Negative	
	zone 1,2,3	zone 1	zone 2,3
Purlin	10.2	- 20.7	- 33.5
Panel	14,4	- 22.9	- 48.3
Fastener	14.4	- 22.9	- 48.3
$A \leq 10$ sq. ft	14.4	- 22.9	- 48.3
$A \geq 100$ sq. ft	10.2	- 20.7	- 33.5

Special case of girt which traverses zones 4 and 5

Width of zone 5
a = 0.1(200) = 20 ft
or = 0.4(36.7) = 14.7 ft (controls)
but > 0.04(200) = 8 ft
> 3 ft

Weighted average design pressure

$$= \frac{14.7(-23.5) + 10.3(-22.1)}{25} = -22.9 \text{ psf}$$

This procedure of weighted average can be used for other components and cladding.

Special Case of Strut Purlin

Strut purlins in the end bay experience combined uplift pressure as a roof component and axial load as part of the MWFRS.

- Component Pressure

  Purlin is located in zone 1 and zone 2
  Width of zone 2, a = 14.7 ft
  Weighted average design pressure

$$= \frac{14.7(-33.4) + 10.3(-20.7)}{25} = -28.2 \text{ psf}$$

- MWFRS Load

  Figure 3.8(a) shows design pressure on end wall with wind parallel to ridge with positive internal pressure (consistent with high uplift on the purlin). Assuming that the end wall is supported at the bottom and at the roof line, the effective axial load on purlin can be determined.

- Combined design load on strut purlin

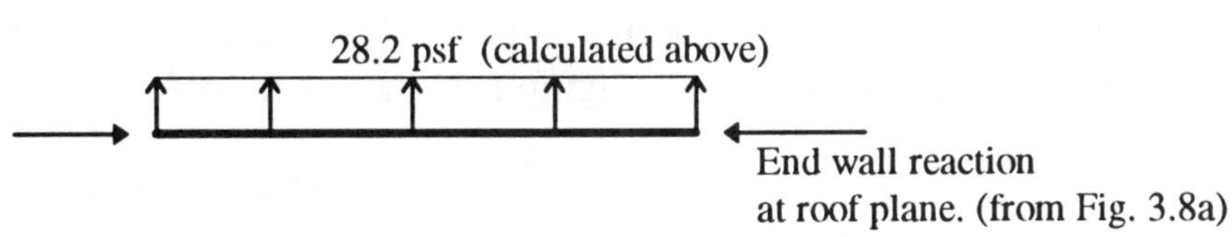

3.4 EXAMPLE 4

This example illustrates the use of the new low-rise building provisions of ASCE 7-95 to determine design pressures for the MWFRS. For this purpose, the building used has the same dimensions as **Example 3**. The design pressures on components and cladding will be the same as **Example 3**. The building data are as follows:

Location:	Memphis, Tennessee
Terrain:	Flat farmland
Dimensions:	200 ft x 250 ft in plan
	Eave height of 20 ft
	Roof Slope 4:12 (18.4 degrees)
Framing:	Rigid frame spans the 200 ft direction
	Rigid frame spacing is 25 ft
	Cross bracing in 250 ft direction
	Openings uniformly distributed

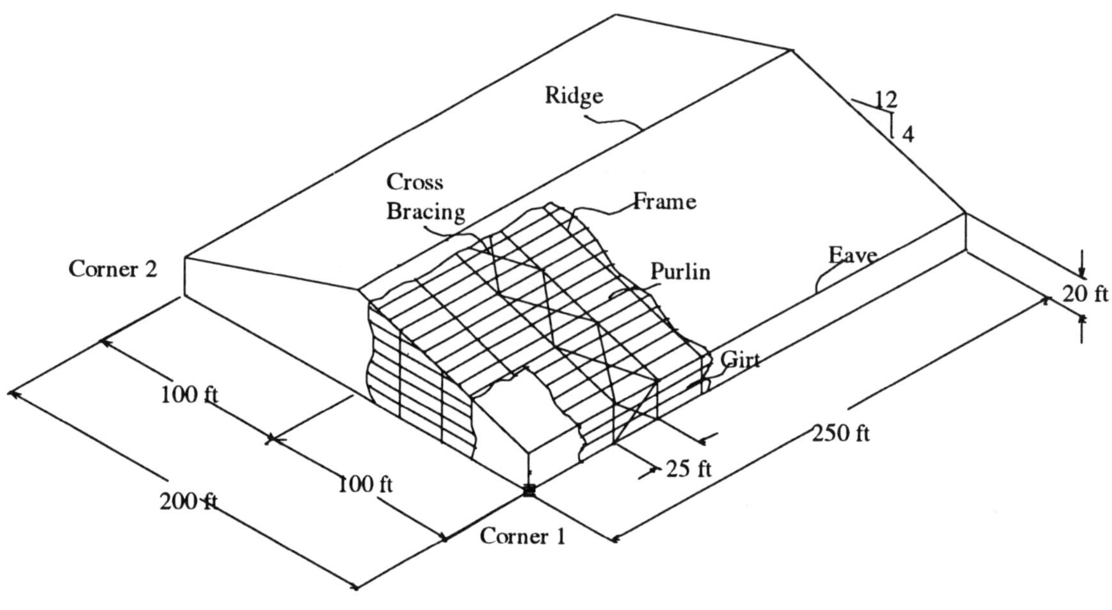

Figure 3.9 Dimensions and Framing of the Low-Rise building of Example 4

<u>Low-Rise Building</u>

The Standard has two requirements for a building to qualify as a Low-Rise Building, (1) mean roof height has to be less or equal to 60 ft, and (2) mean roof height does not exceed least horizontal dimension (Section 6.2). A building with these dimensions qualifies to be a low-rise building and external pressure coefficients of Figure 6-4 may be used.

Exposure, Building Classification and Basic Wind Speed

Same as Example 3: Exposure C
Category II
V = 90 mph

Velocity Pressures

The low-rise building provisions for MWFRS in the Standard use the velocity pressure at mean roof height, h, for calculation of all external and internal pressures including the windward wall.

Mean roof height h = 36.7 ft

The velocity pressures are computed using:

$$q_h = 0.00256 K_h K_{zt} V^2 I \text{ psf} \qquad \text{(Eq. 6-1)}$$

where:

q_h = velocity pressure at mean roof height, h

K_h = 1.02 for Exposure C

K_{zt} = 1.0 topographic factor

V = 90 mph basic wind speed

I = 1.0 for Category II

$$q_h = 0.00256(1.02)(1.0)(90)^2(1.0) = 21.2 \text{ psf}$$

Design Pressures for the MWFRS

The equation for the MWFRS for low-rise buildings in Table 6-1 is:

$$p = q_h[(GC_{pf}) - (GC_{pi})]$$

where:

q_h is the velocity pressure at mean roof height associated with Exposure C

(GC_{pf}) is the external pressure coefficients from Figure 6-4

(GC_{pi}) is the internal pressure coefficient from Table 6-4

To obtain the critical wind actions, including torsion, for design of the MWFRS, the two separate loading cases, Case A and Case B, indicated in Figure 6-4 must be applied at each corner (See Figure C6-2). For each of these cases, both positive and negative internal pressures must be considered, resulting in a total of 16 separate loading conditions. However, if the building is symmetrical, the number of separate loading conditions will be reduced to 8 (2 loading cases, 2 windward corners, and 2 internal pressures).

In applying the loads of Case A and Case B to corner 2 shown in Figure 3.9 (building rotated 90 degrees), zones 2 and 3 are assumed to be separated by an imaginary ridge line oriented normal to and bisecting the actual ridge line. Additionally, the roof angle, θ, is assumed to be zero degrees when selecting the values of GC_{pf} for Case A loading at corner 2 (see Figure C6-2).

External (GC_{pf})

The roof and wall coefficients are functions of the roof slope, θ. There are 8 building surfaces defined for Case A: 4 interior surfaces and 4 end zone surfaces. For Case B, there are 12 surfaces identified: 6 interior surfaces and 6 end zone surfaces.

Width of end zone surface: 2a = 2(0.1)(200) = 40 ft
or = 2(0.4)(36.7) = 29.4 ft (controls)
but > 2(0.04)(200) = 16 ft
> 2 (3) = 6 ft

Corner 1: Case A, GC_{pf} (θ = 18.43 degrees)

θ	Surface							
	1	2	3	4	1E	2E	3E	4E
0 - 5	0.40	- 0.69	- 0.37	- 0.29	0.61	- 1.07	- 0.53	- 0.43
18.4*	0.52	- 0.69	- 0.47	- 0.42	0.78	- 1.07	- 0.67	- 0.62
20	0.53	- 0.69	- 0.48	- 0.43	0.80	- 1.07	- 0.69	- 0.64

* By linear interpolation

Corner 1 and 2: Case B, GC_{pf}

Location	Surface					
	1	2	3	4	5	6
Interior	- 0.45	- 0.69	- 0.37	- 0.45	0.40	- 0.29
Edge*	- 0.48	- 1.07	- 0.53	- 0.48	0.61	- 0.43

*Zone 1E, 2E, 3E, 4E, 5E,and 6E in Figure 6-4

Corner 2: Case A, GC_{pf} (θ = 0 degrees)

θ	Surface							
	1	2	3	4	1E	2E	3E	4E
0 - 5	0.40	- 0.69	- 0.37	- 0.29	0.61	- 1.07	- 0.53	- 0.43

Internal (GC_{pi})

Openings are assumed to be evenly distributed in the walls. Since Memphis, TN is not in a hurricane-prone region the internal pressure coefficient is:

$$(GC_{pi}) = \pm 0.18$$

Design Pressures, psf

Typical calculations for design pressures are as follows:

Corner 1, Case A: Surface 1

$$p = 21.2[(0.52) - (\pm 0.18)]$$

p = 14.8 psf with (-) internal pressure

p = 7.2 psf with (+) internal pressure

Corner 1, Case A: Surface 2E

$$p = 21.2[(-1.07) - (\pm 0.18)]$$

p = - 18.8 psf with (-) internal pressure

p = - 26.5 psf with (+) internal pressure

Design Pressures, Corner 1: Case A

Surface	(GC_{pf})	Design Pressure, psf	
		$(+GC_{pi})$	$(-GC_{pi})$
1	0.52	7.2	14.8
2	-0.69	- 18.4	- 10.8
3	-0.47	- 13.7	- 6.1
4	-0.42	- 12.7	- 5.1
1E	0.78	12.7	20.3
2E	-1.07	- 26.5	- 18.8
3E	-0.67	- 18.0	- 10.4
4E	-0.62	- 16.9	- 9.3

Design Pressures, Corner 1 and 2: Case B

Surface	(GC_{pf})	Design Pressure, psf	
		$(+GC_{pi})$	$(-GC_{pi})$
1	- 0.45	- 13.3	- 5.7
2	- 0.69	- 18.4	- 10.8
3	- 0.37	- 11.6	- 4.0
4	- 0.45	- 13.3	- 5.7
5	0.40	4.7	12.3
6	- 0.29	- 9.9	- 2.3
1E	- 0.48	- 14.0	- 6.3
2E	- 1.07	- 26.5	- 18.8
3E	- 0.53	- 15.0	- 7.4
4E	- 0.48	- 14.0	- 6.3
5E	0.61	9.1	16.7
6E	- 0.43	- 12.9	- 5.3

Design Pressures, Corner 2: Case A

Surface	(GC_{pf})	Design Pressure, psf	
		($+GC_{pi}$)	($-GC_{pi}$)
1	0.40	4.7	12.3
2	- 0.69	- 18.4	- 10.8
3	- 0.37	- 11.6	- 4.0
4	- 0.29	- 9.9	- 2.3
1E	0.61	9.1	16.7
2E	- 1.07	- 26.5	- 18.8
3E	- 0.53	- 15.0	- 7.4
4E	- 0.43	- 12.9	- 5.3

Application of Pressures on Building Surfaces 2 and 3

Footnote 4(a) of Figure 6-4 states that the roof pressure coefficient GC_{pf}, when negative in zone 2, shall be applied in zone 2 for a distance from the edge of the roof equal to 0.5 times the horizontal dimension of the building measured perpendicular to the eave line or 2.5h, whichever is less; the remainder of zone 2 extending to the ridge line shall use the pressure coefficient GC_{pf} for zone 3.

0.5(200) = 100 ft
2.5h = (2.5)(36.7) = 92 ft (controls)

Therefore zone 3 applies over a distance of 105 - 92 = 13 ft in what is normally considered to be zone 2 (adjacent to ridge line).

Loading Cases

Because the building is symmetrical, the eight loading cases provide all the combinations, provided the design is accomplished applying loads for each of the four corners. These load combinations shown in Figures 3.10 through 3.17 are to be used to design the rigid frame and the cross bracing.

The design pressures on components and cladding are the same as shown in Example 3.

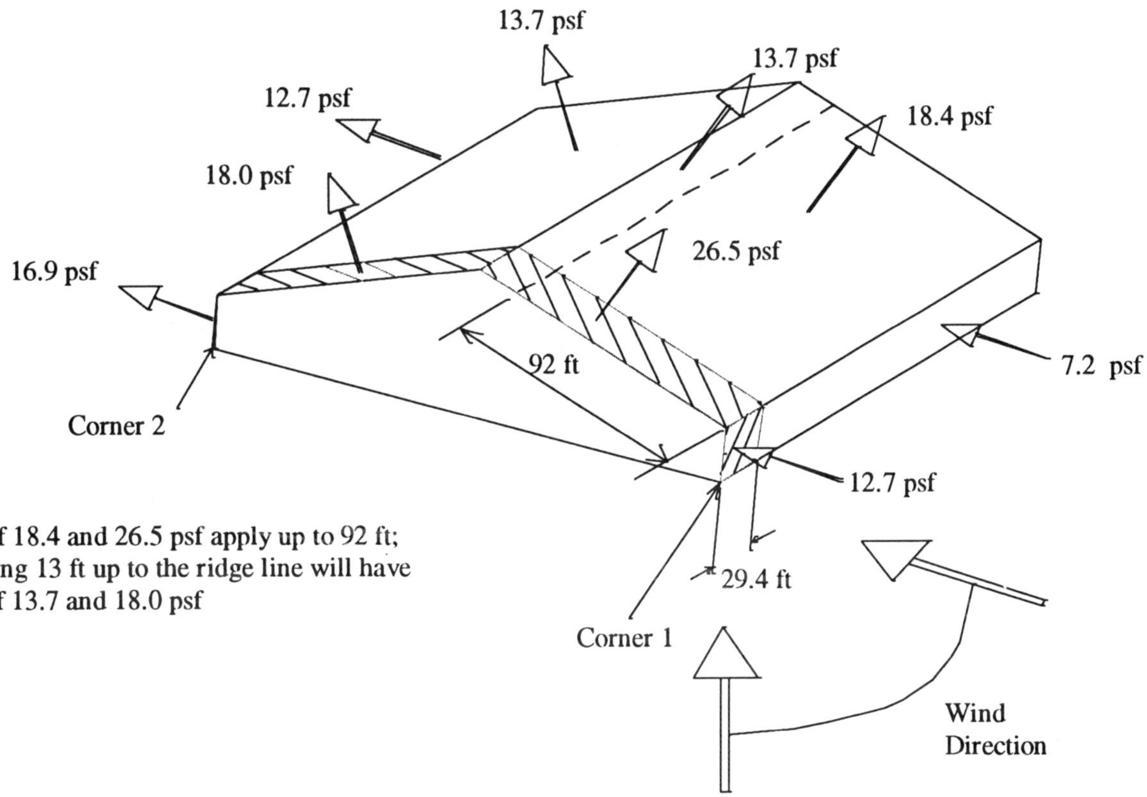

Note:
Pressures of 18.4 and 26.5 psf apply up to 92 ft; the remaining 13 ft up to the ridge line will have pressures of 13.7 and 18.0 psf

Figure 3.10 Design Pressures for Case A at Corner 1 with Positive Internal Pressure

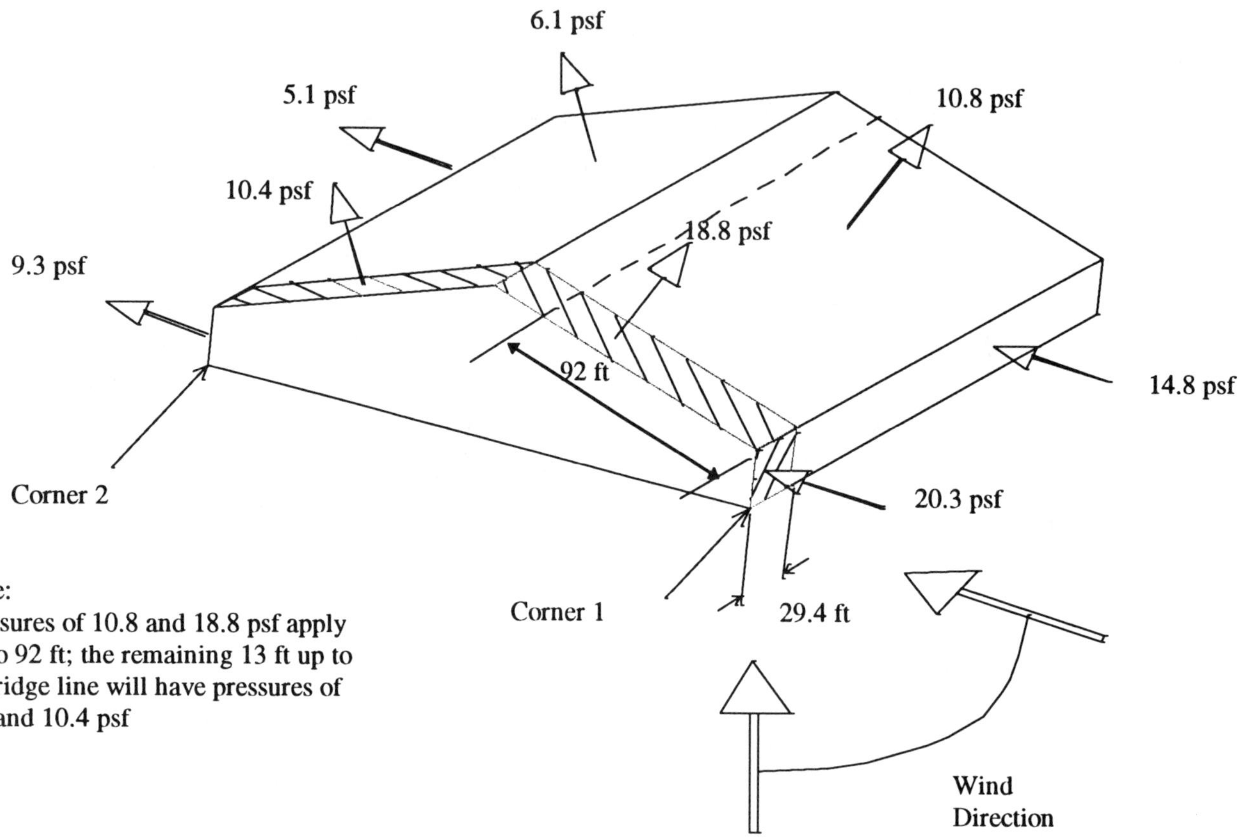

Note:
Pressures of 10.8 and 18.8 psf apply up to 92 ft; the remaining 13 ft up to the ridge line will have pressures of 6.1 and 10.4 psf

Figure 3.11 Design Pressures for Case A at Corner 1 with Negative Internal Pressure

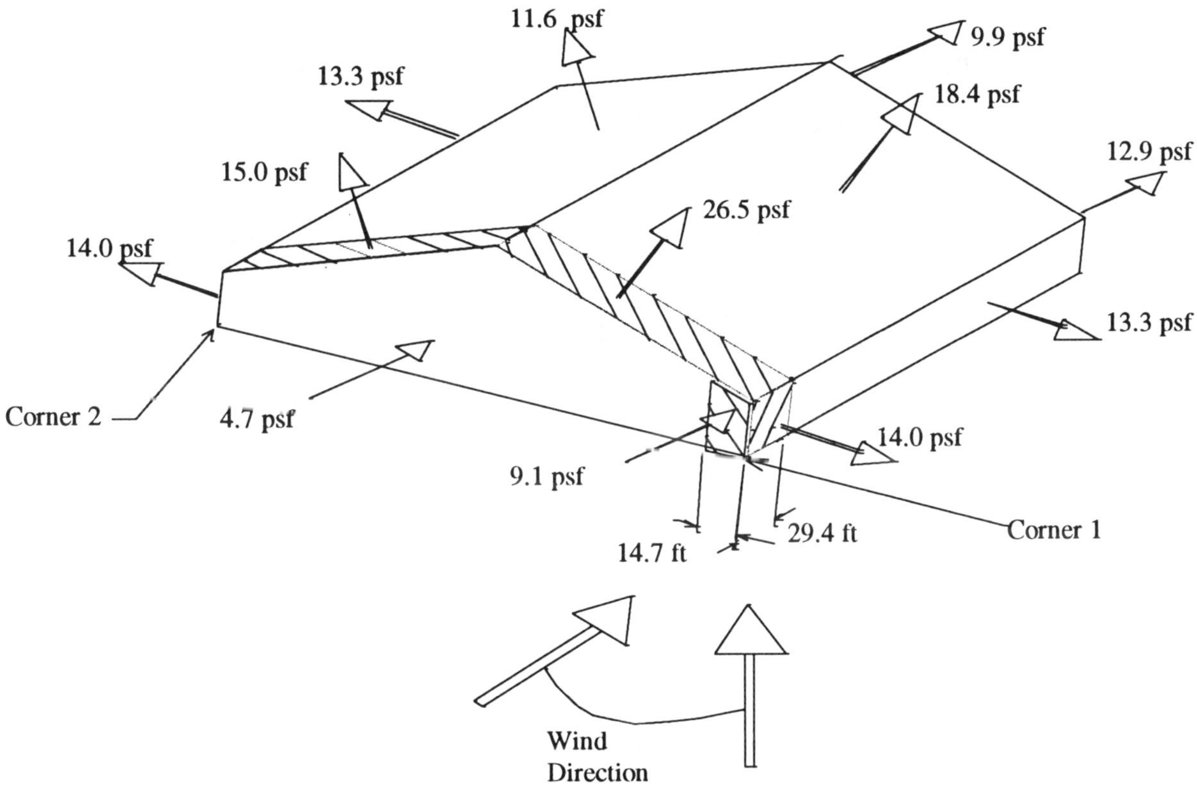

Figure 3.12 Design Pressures for Case B at Corner 1 with Positive Internal Pressure

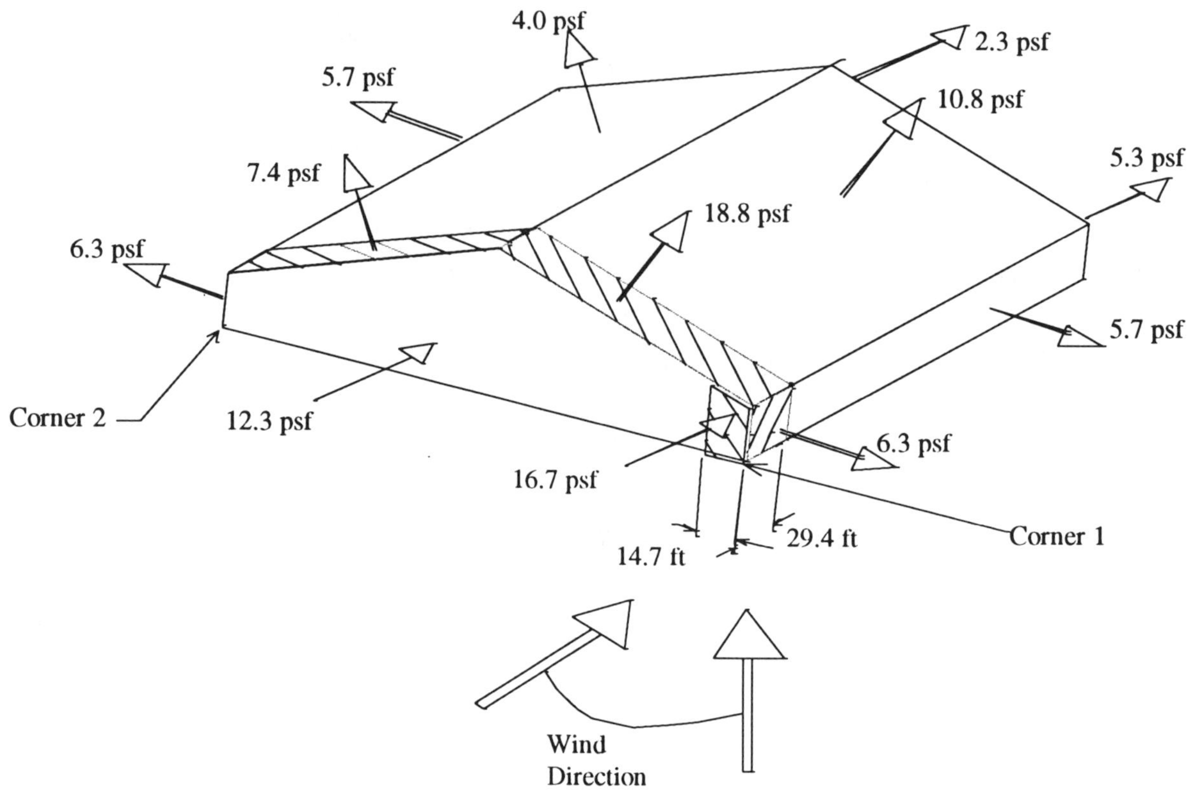

Figure 3.13 Design Pressures for Case B at Corner 1 with Negative Internal Pressure

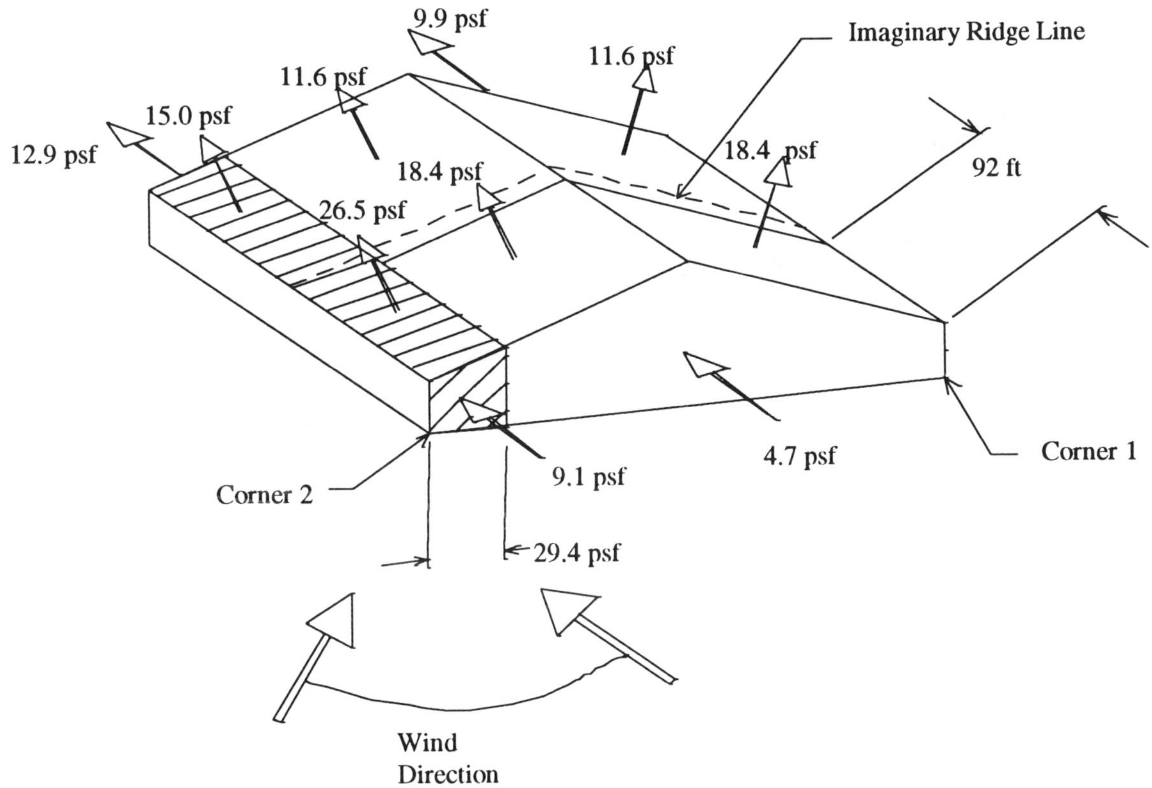

Figure 3.14 Design Pressures for Case A at Corner 2 with Positive Internal Pressure

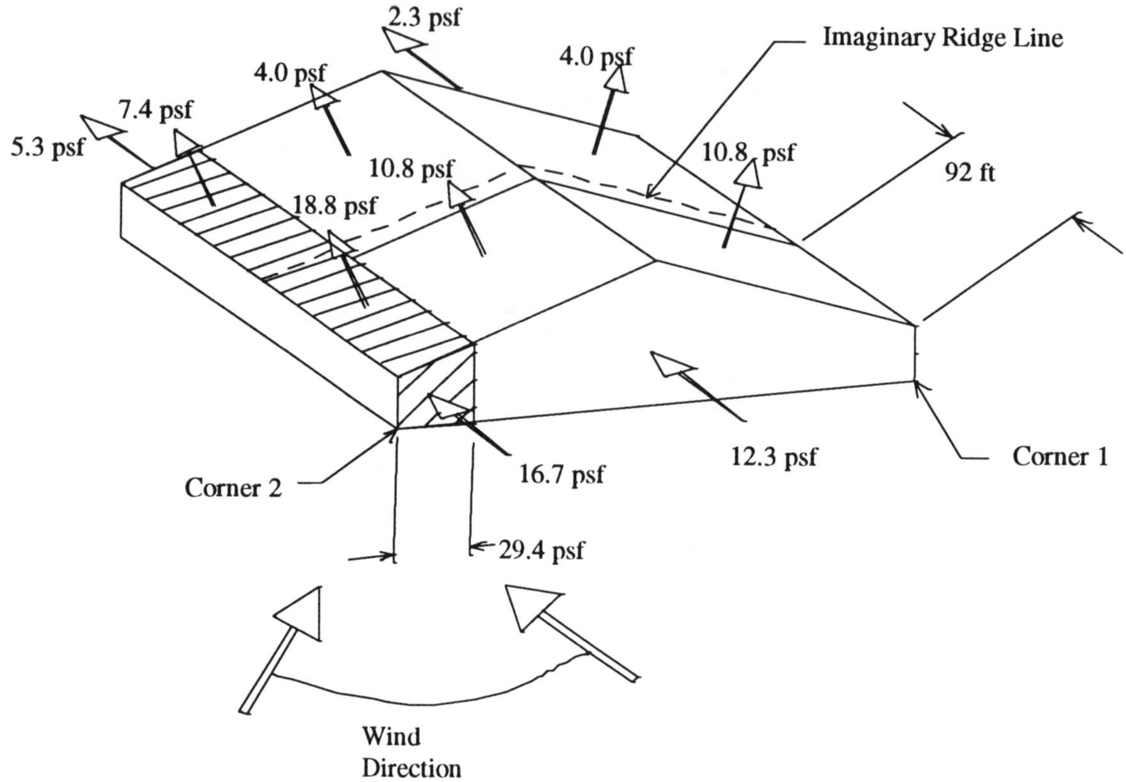

Figure 3.15 Design Pressures for Case A at Corner 2 with Negative Internal Pressure

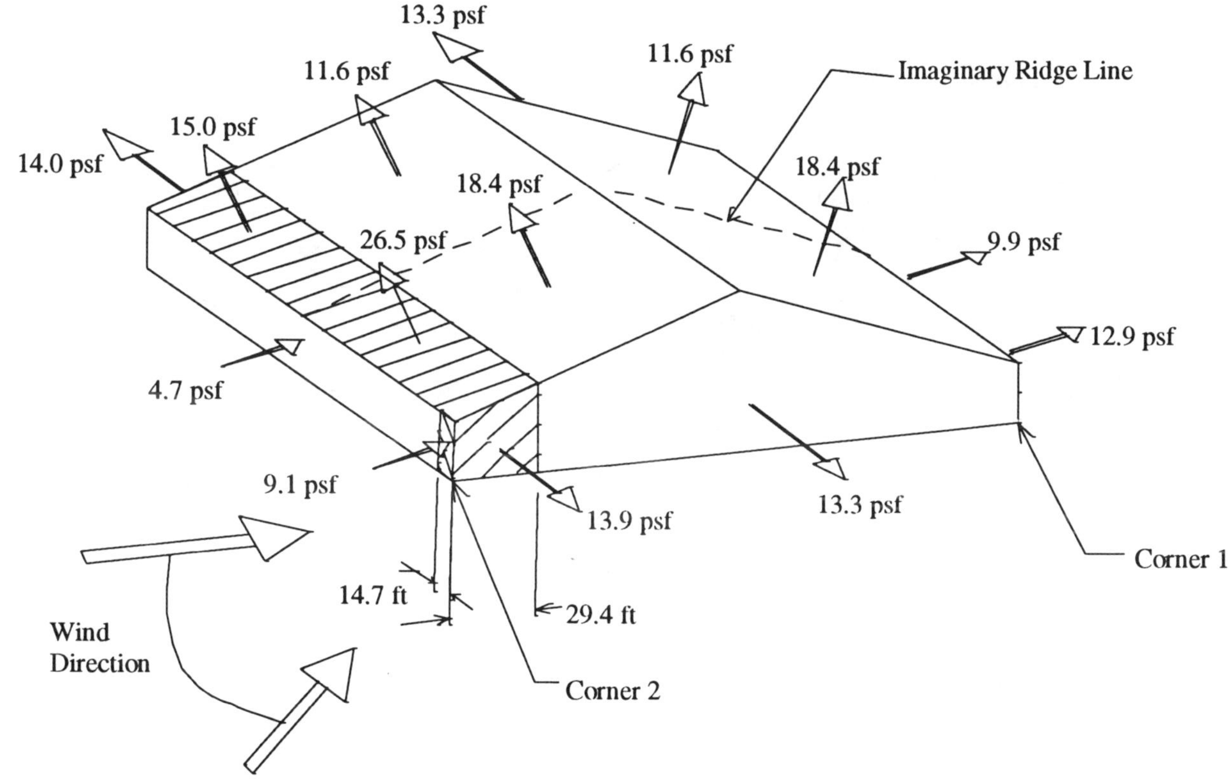

Figure 3.16 Design Pressures for Case B at Corner 2 with Positive Internal Pressure

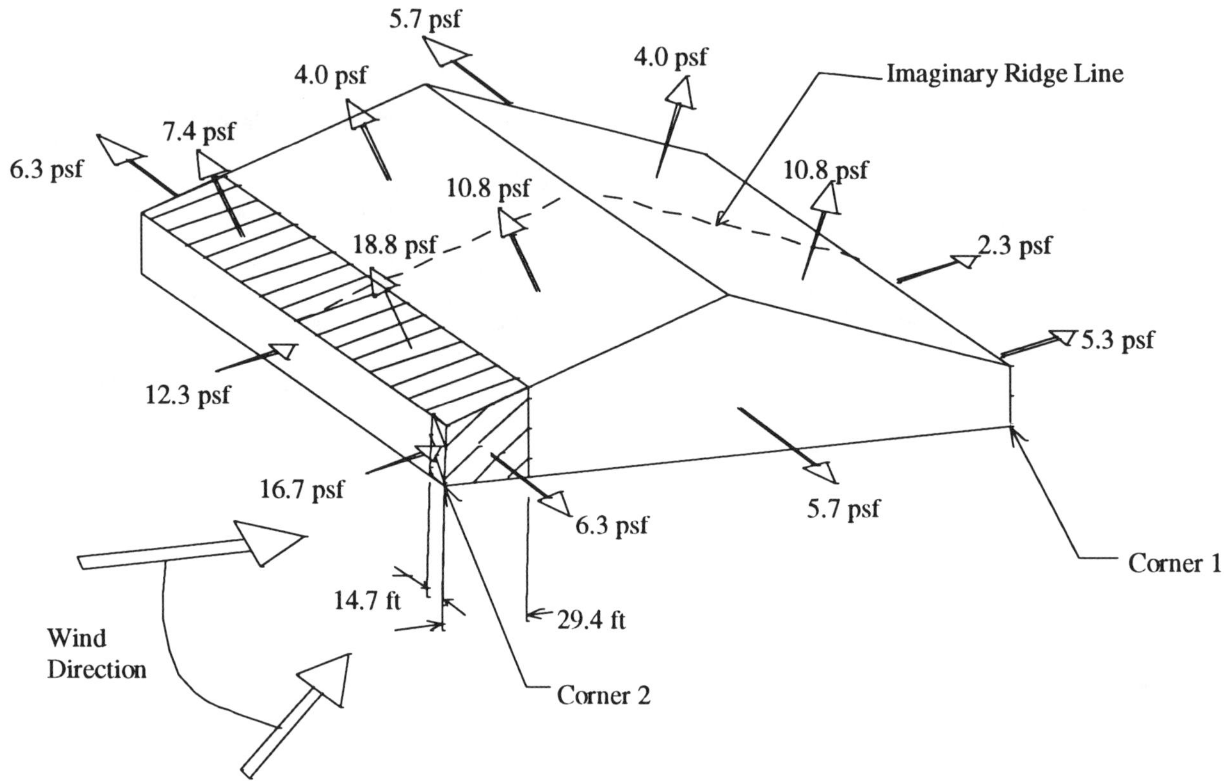

Figure 3.17 Design Pressures for Case B at Corner 2 with Negative Internal Pressure

<u>Comment: Exposure B Terrain</u>

If the building is located in terrain representative of Exposure B in <u>all</u> directions, the low-rise MWFRS design pressures and the component and cladding pressures can be reduced by 15%; see note 6 in Figure 6-4, note 10 in Figure 6-5 and note 4 in Table 6-4.

3.5 EXAMPLE 5

In this example, design pressures for a typical retail store strip-mall are determined. The building data are as follows:

Location:	Boston, Massachusetts; about seven miles from shoreline
Topography:	Homogeneous
Terrain:	Suburban
Dimensions:	40 ft x 80 ft in plan
	Monoslope roof with slope of 14 degrees and overhang of 7 ft in plan
	Wall heights are 15 ft in front and 25 ft in rear
Framing:	Walls of CMU on all sides supported at top and bottom; steel framing in front to hold window glass and doors
	Roof joists span 41.2 ft with 7.2 ft overhang spaced at 5 ft on center
Cladding:	Glass and door sizes vary; glazing is not debris impact resistant
	Roof panels are 2 ft wide and 20 ft long

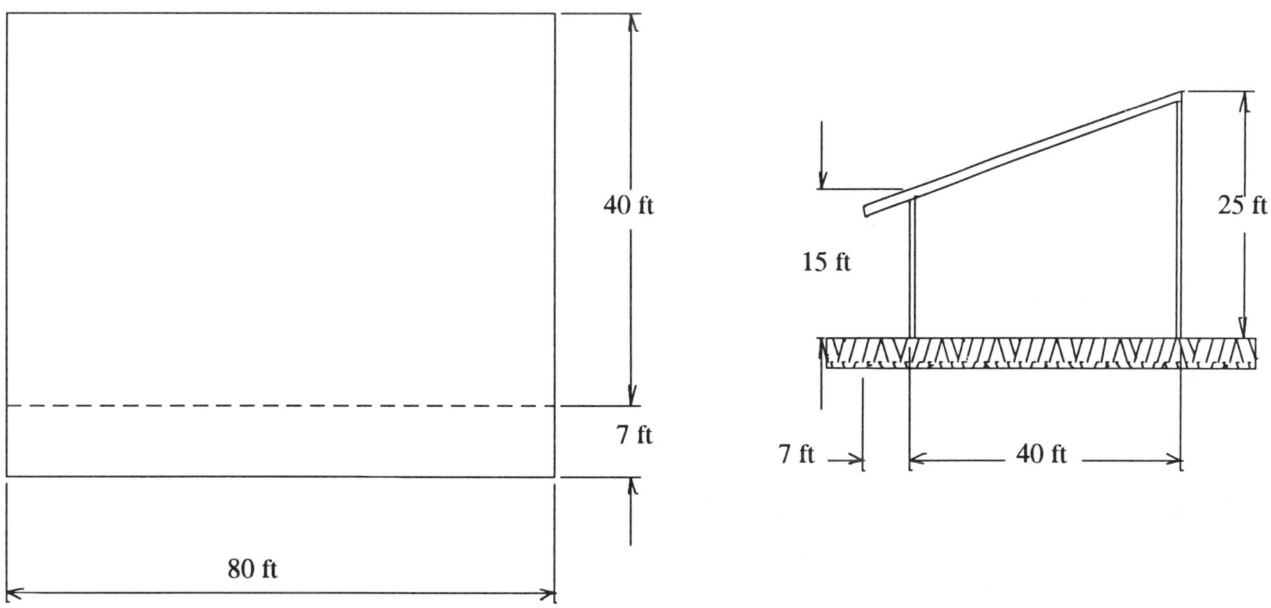

Figure 3.18 Dimensions of the Retail Store Strip-mall

<u>Exposure and Building Classification</u>

The building is in suburban terrain; use Exposure B.

The building is not an essential facility or likely to be occupied by more than 300 persons at one time. Use Category II.

Basic Wind Speed

Wind speed contour of 110 mph traverses over Boston, Massachusetts. Use a basic wind speed = 110 mph.

Velocity Pressures

The velocity pressures are computed using:

$$q_z = 0.00256 K_z K_{zt} V^2 I \text{ psf} \qquad \text{(Eq. 6-1)}$$

Where, K_z is obtained from Table 6-3, K_{zt} = 1.0 (homogeneous terrain), I = 1.0 for category II building, and the basic wind speed is V = 110 mph.

The velocity pressures q are based on Exposure B for MWFRS; however the design of components and cladding for buildings with mean roof height, h less than 60 ft requires use of q_h based on Exposure C (Table 6-1). Mean roof height h = 20 ft.

Velocity Pressures, q_z and q_h, psf

Height, ft	K_z		q_z , q_h	
	Exp. B	Exp. C	Exp. B	Exp. C
0-15	0.57	-	17.7	-
h = 20	0.62	0.90	19.2	27.9
25	0.66	-	20.4	-

Design Pressures for MWFRS

The equation for buildings of all heights in Table 6-1 is:

$$p = qGC_p - q_h(GC_{pi})$$

Where, q is q_z for windward wall and q_h for leeward wall, side walls and roof, G is determined from Section 6.6.1, values of C_p are obtained from Figure 6-3 and values of (GC_{pi}) are obtained from Table 6-4.

Note:

The provisions of the standard permit use of external pressure coefficients (GC_{pf}) given in Figure 6-4; however the values in Figure 6-4 were obtained from wind tunnel studies of gable frame buildings. Their use for a monoslope roof requires considerable judgment. Example 4 with a gable frame illustrates use of the pressure coefficients of Figure 6-4, and the commentary in the standard gives the background for (GC_{pf}) values.

Gust Effect Factor, G

The gust effect factor for non-flexible buildings in Exposure B is:

$$G = 0.8$$

Wall External Pressure Coefficients, C_p

The values for the leeward wall depend on L/B; they are different for the two directions, (1) wind parallel to roof slope, and (2) wind normal to roof slope.

Wall Pressure Coefficients, C_p

Surface	Wind Direction	L/B	Cp
Leeward Wall	ll to roof slope	0.5	- 0.5
Leeward Wall	⊥ to roof slope	2.0	- 0.3
Windward Wall	-	-	0.8
Side Walls	-	-	- 0.7

Roof External Pressure Coefficients, C_p

Since the building has a monoslope roof ,the roof surface for wind directed parallel to the slope may be a windward or a leeward surface. The value of h/L = 0.5 in this case.

When wind is normal to the roof slope, angle $\theta = 0$ and h/L = 0.25.

Roof Pressure Coefficients, C_p

Wind Direction	h/L	θ, degree	C_p
ll to roof slope	0.5	14	-0.74 as windward slope
ll to roof slope	0.5	14	-0.50 as leeward slope
⊥ to roof slope	0.25	0	-0.90 (0-20 ft)*
			-0.50 (20-40 ft)
			-0.30 (40-80 ft)

* Distance from the windward edge of the roof

For the overhang, Section 6.7.2.1 requires $C_p = 0.8$.

Internal Pressure Coefficients, (GC_{pi})

The building is located in a hurricane-prone region. The basic wind speed is 110 mph and the glazing is not designed to resist wind-borne debris impact. The building satisfies both conditions in Table 6-4. Irrespective of the openings in the walls and the roof, the internal pressure coefficients are:

$$(GC_{pi}) = +0.80 \text{ and}$$
$$(GC_{pi}) = -0.30$$

Typical Calculations of Design Pressures for MWFRS (wind parallel to slope with 15 ft windward wall)

Pressure on Leeward wall

$p = q_h GC_p - q_h(\pm GC_{pi})$

$= 19.2(0.8)(-0.5) - (19.2)(0.8)$

$= -23.0$ psf with positive internal pressure

and

$= 19.2\,(0.8)(-0.5) - (19.2)\,(-0.3)$

$= -1.9$ psf with negative internal pressure

Pressure on Overhang Top Surface

$p = q_h GC_p$

$= 19.2(0.8)(-0.74)$

$= -11.4$ psf

Pressure on Overhang Bottom Surface
(same as windward wall external pressure)

$p = q_z GC_p$

$= 17.7(0.8)(0.8)$

$= 11.3$ psf

Design Pressures for MWFRS: Wind Parallel to Roof Slope

Wind Direction	Surface	z ft	q_z , q_h psf	External C_p*	Design Pressure, psf (+GC_{pi})	Design Pressure, psf (-GC_{pi})
Windward Wall 15 ft	Windward wall	0-15	17.7	0.80	-4.1	17.1
	Leeward wall	0-25	19.2	-0.50	-23.0	-1.9
	Side wall	all	19.2	-0.70	-26.1	-5.0
	Roof	-	19.2	-0.74	-26.7	-5.6
	Overhang top	-	19.2	-0.74	-11.4**	-11.4**
	Overhang bot	-	17.7	0.80	11.3**	11.3**
Windward Wall 25 ft	Windward wall	0-15	17.7	0.80	-4.1	17.1
		15-20	19.2	0.80	-3.1	18.1
		20-25	20.4	0.80	-2.3	18.8
	Leeward wall	all	19.2	-0.50	-23.0	-1.9
	Side wall	all	19.2	-0.7	-26.1	-5.0
	Roof	-	19.2	-0.5	-23.0	-1.9
	Overhang top	-	19.2	-0.5	-7.7**	-7.7**
	Overhang bot	-	-	-	0.0**	0.0**

* External pressure calculations include G = 0.80

** Overhang pressures are not affected by internal pressures

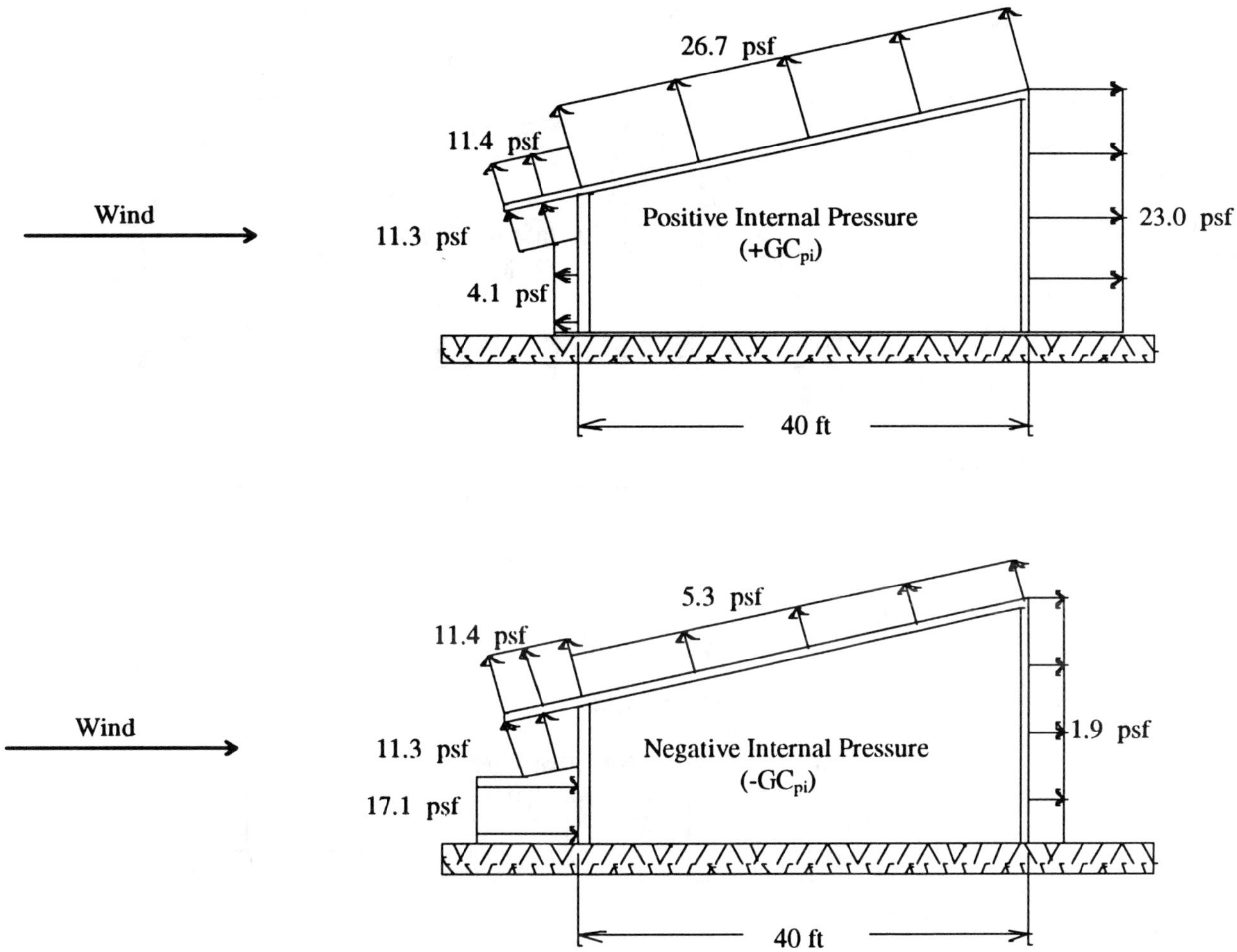

Figure 3.19 Design Pressures for MWRFS; Wind Parallel to Roof Slope

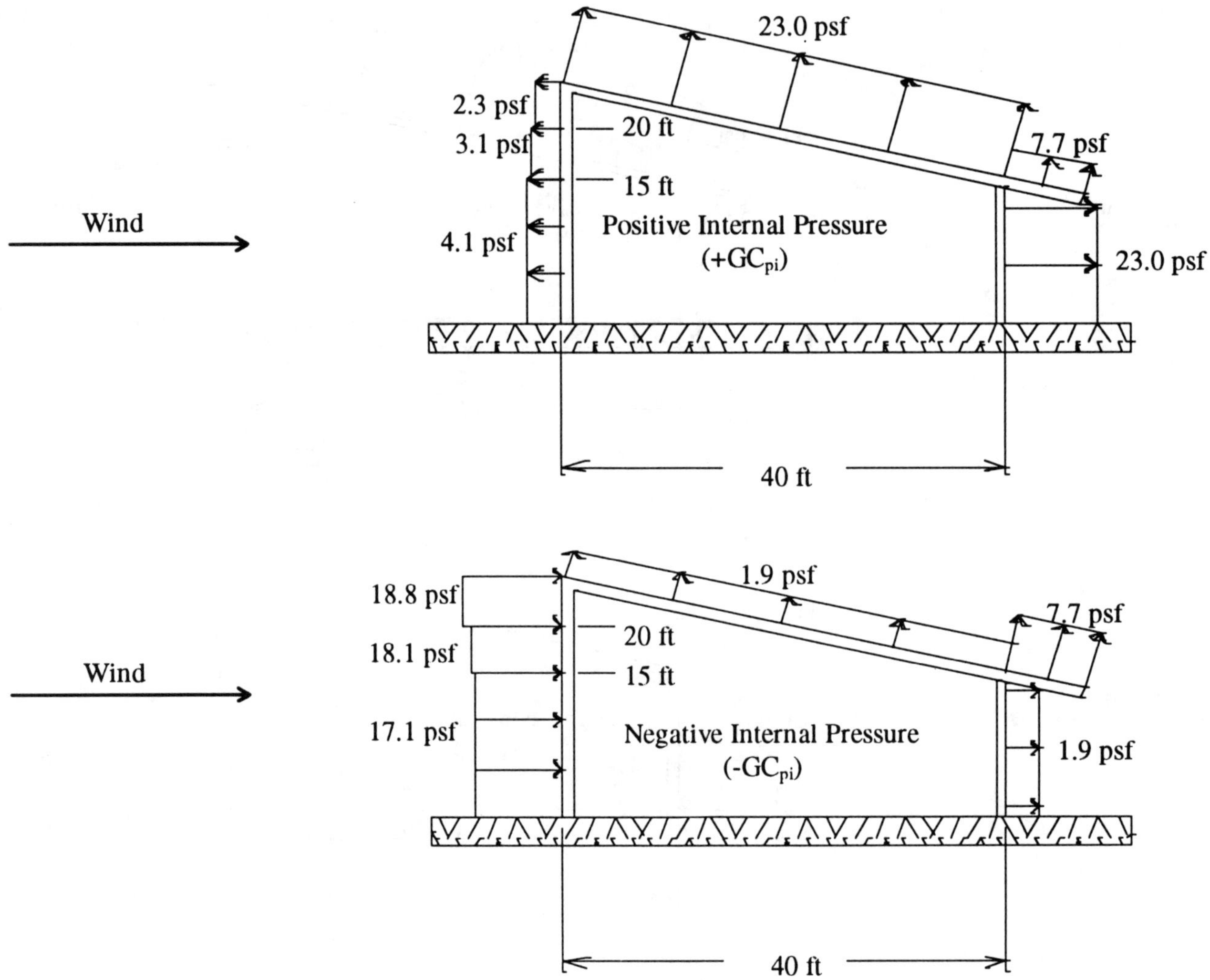

Figure 3.20 Design Pressures for MWRFS, Wind Parallel to Roof Slope

Design Pressures for MWFRS: Wind Normal to Roof Slope

Surface	z or Distance*** ft	q_z , q_h* psf	C_p	Design Pressure, psf ($+GC_{pi}$)**	Design Pressure, psf ($-GC_{pi}$)**
Windward wall	0-15	17.7	0.8	-4.1	17.1
	15-20	19.2	0.8	-3.1	18.1
	20-25	20.4	0.8	-2.3	18.8
Leeward wall	all	19.2	-0.3	-20.0	1.2
Side wall	all	19.2	-0.7	-26.1	-5.0
Roof ****	0-20	19.2	-0.9	-29.2	-8.1
	20-40	19.2	-0.5	-23.0	-1.9
	40-80	19.2	-0.3	-20.0	1.2

* External pressure calculations include G = 0.80

** Internal pressure is associated with q_h = 19.2 psf

*** Distance along roof is from leading windward edge

****Pressure on overhang is only external pressure

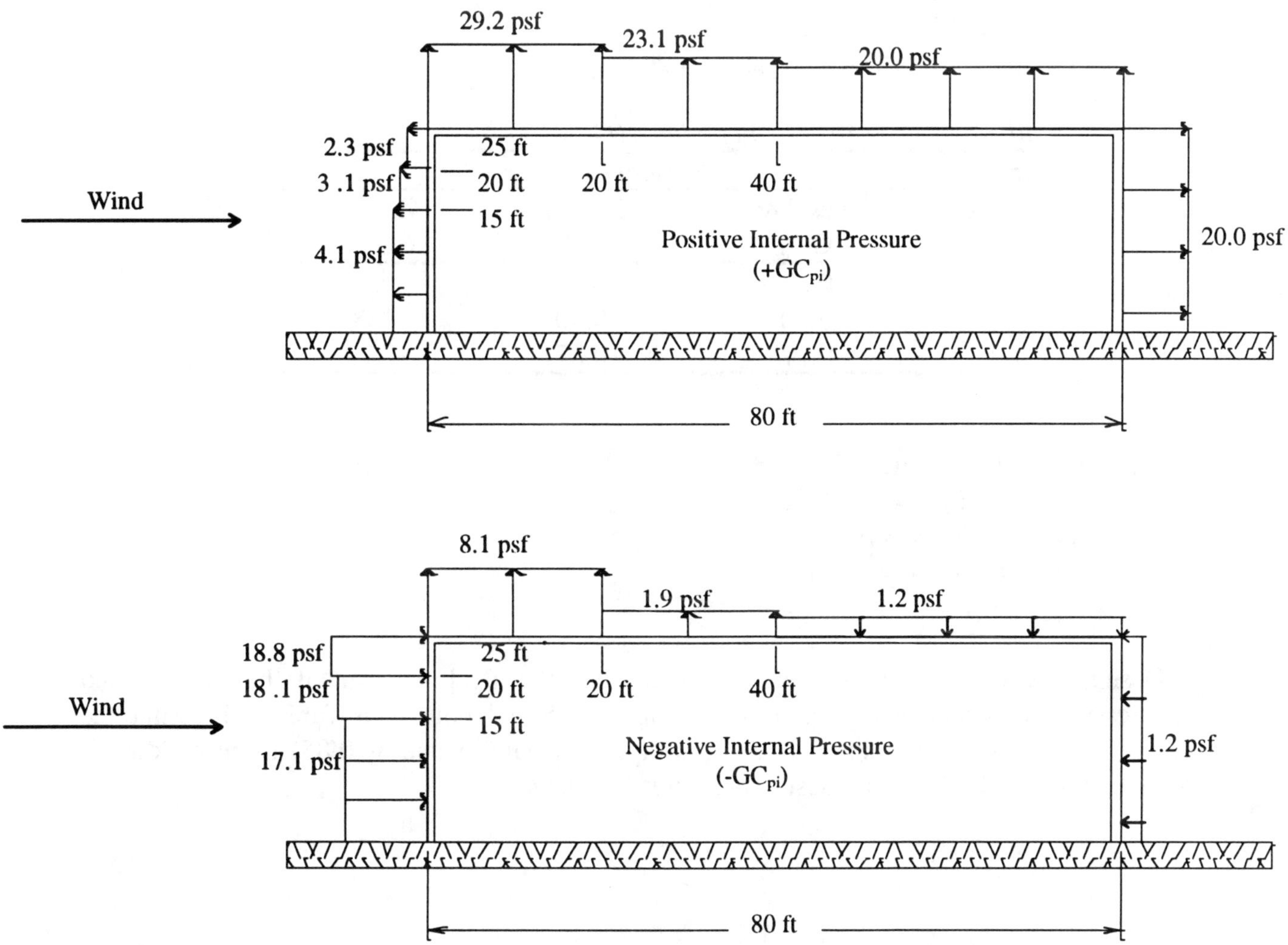

Figure 3.21 Design Pressures for MWFRS, Wind Perpendicular to Roof Slope

Design Pressures for Components and Cladding

Design pressure equation for components and cladding for building with mean roof height $h \leq 60$ ft is given in Table 6-1.

$$p = q_h[(GC_p) - (GC_{pi})]$$

where q_h is velocity pressure at mean roof height associated with Exposure C (q_h = 27.9 psf previously determined); (GC_p) are external pressure coefficients from Figure 6-5 and Figure 6-7A; (GC_{pi}) = +0.80 and -0.30 (previously determined) from Table 6-4.

Wall Design Pressures

Since the CMU walls are supported on the top and bottom, the effective wind area will depend on the span length.

Effective wind area:

for span of 15 ft, A = 15(15/3) = 75 sq. ft
for span of 20 ft, A = 20(20/3) = 133 sq. ft
for span of 25 ft, A = 25(25/3) = 208 sq. ft

Wall External Pressure Coefficients (GC_p)

Area A	Pressure Coefficients		
sq. ft	Zones 4 & 5 (GC_p)	Zone 4 ($-GC_p$)	Zone 5 ($-GC_p$)
75	0.85	-0.95	-1.09
133	0.80	-0.90	-1.00
208	0.77	-0.87	-0.93

Width of zone 5 (Figure 6-5)

a = 0.1(40) = 4 ft (controls)
or = 0.4(20) = 8 ft
but > 0.04(40) = 1.6 ft
> 3 ft

Design pressures are the critical pressures when the algebraic sum of the external and internal pressures is a maximum. Note 10 in Figure 6-5 and note 4 in Table 6-4 permit the reduction of the calculated pressure by 15% (multiply calculated value by 0.85) if the building is sited within Exposure B. The wall design pressures are shown.

Typical calculations for Design Pressures for 15 ft Wall, Zone 4

$$p = q_h [(GC_p) - (\pm GC_{pi})] (0.85)$$
$$= 27.9 [(0.85) - (-0.3)] (0.85)$$
$$= 27.3 \text{ psf}$$
and
$$= 27.9 [(-0.95) - (0.8)] (0.85)$$
$$= -41.5 \text{ psf}$$

Wall Design Pressures, psf

Wall Height ft	Design Pressures, psf Zone 4 & 5 positive	Zone 4 Negative	Zone 5 Negative
15	27.3	-41.5	-44.8
20	26.1	-40.3	-42.7
25	25.4	-39.6	-41.0

Notes: q_h = 27.9 psf
pressure values include 0.85 for terrain exp. B

The CMU walls should be designed for pressures determined for zone 4 and 5 using appropriate tributary areas.

The design pressures for doors and glazing can be assessed by using appropriate pressure coefficients associated with their effective wind areas.

Roof Design Pressures

Effective wind area:

Roof joist, A = (41.2+7.2)(5) = 242 sq. ft
or = (48.4)(48.4/3) = 781 sq. ft (controls)

Roof panel, A = (5)(2) = 10 sq. ft (controls)
or = (5)(5/3) = 8.3 sq. ft

Even though the effective wind area of the roof joist is more than 700 sq. ft, its external pressure coefficients (GC_p) are determined on the basis of components and cladding. The note in Table 6-1, in which provisions for MWFRS may be used for a major component is valid only when the tributary area is greater than 700 sq. ft. The tributary area for the roof joist is 242 sq. ft.

Section 6.7.2.2 requires that pressure coefficients for roof overhangs be obtained from Figure 6-5B. The zones for roof overhangs in Figure 6-5B are different from the zones for a monoslope roof in Figure 6-7A.

Roof External Pressure Coefficients (GC_p), θ = 14 degree

Component	Area A sq. ft	Zones 1,2,3 (GC_p)	Zone 1 ($-GC_p$)	Zone 2 ($-GC_p$)	Zone 3 ($-GC_p$)
		Pressure Coefficient, Figure 6-7A			
Joist	781	0.3	-1.1	-1.2	-2.0
Panel	10	0.4	-1.3	-1.6	-2.9
		Pressure Coefficient, Figure 6-5B			
		Zones 1,2,3 (GC_p)	Zone 1 ($-GC_p$)	Zone 2 ($-GC_p$)	Zone 3 ($-GC_p$)
Joist	781	0.3	-0.8	-2.2	-2.5
Panel	10	0.5	-0.9	-2.2	-3.7

Width of zone distance a

a = 0.1(40) = 4 ft (controls)
or = 0.4(20) = 8 ft
> 0.04(40) = 1.6 ft
> 3 ft

The widths and lengths of zones 2 and 3 for a monoslope roof are shown in Figure 6-7A (they vary from a to 4a), and for overhangs in Figure 6-5B.

Similar to the determination of design pressures for walls, the critical design pressures for roofs are the algebraic sum of the external and internal pressures. In addition, the calculated pressures are multiplied by 0.85 because the building is sited in Exposure B. The design pressures for overhang areas are based on pressure coefficients obtained from Figure 6-5B.

Typical calculations for Joist Pressures

Zone 2:

$$p = q_h [(GC_p) - (\pm GC_{pi})] (0.85)$$
$$= 27.9 [(0.3) - (-0.3)] (0.85)$$
$$= 14.2 \text{ psf}$$

and

$$= 27.9 [(-1.2) - (0.8)] (0.85)$$
$$= -47.4 \text{ psf}$$

Roof Design Pressures, psf

Component	Design Pressures, psf			
	Zones 1,2,3* positive	Zone 1 Negative	Zone 2 Negative	Zone 3 Negative
Joist	14.2	-45.0	-47.4	-66.4
Joist Overhang	10.0**	-19.0	-52.1	-59.2
Panel	16.6	-49.8	-56.9	-87.7
Panel in Overhang	11.8	-21.3	-52.1	-87.7

Note: q_h = 27.9 psf; design pressures include 0.85 for exposure B
* zones for overhang are in accordance with Figure 6-5B
** Section 6.4.1.2 requires minimum of 10 psf

Zones for the monoslope roof and for overhang are shown in Figure 3.22. The panels should be designed for the pressures indicated.

Roof joist design pressures need careful interpretation. The high pressures in corner or eave areas do not occur simultaneously at both ends. Two loading cases on two joists are as follows and shown in Figure 3.22.

- Joist 1, loading 1: zones 2 and 3 for roof and zone 2 for overhang
- Joist 1, loading 2: zones 2 for roof and zones 2 and 3 for overhang
- Joist 2, loading 3: zones 1 and 2 for roof and zone 1 for overhang
- Joist 2, loading 4: zone 1 for roof and zones 1 and 2 for overhang

For simplicity, only one zone is used for overhang pressures in Figure 3.22.

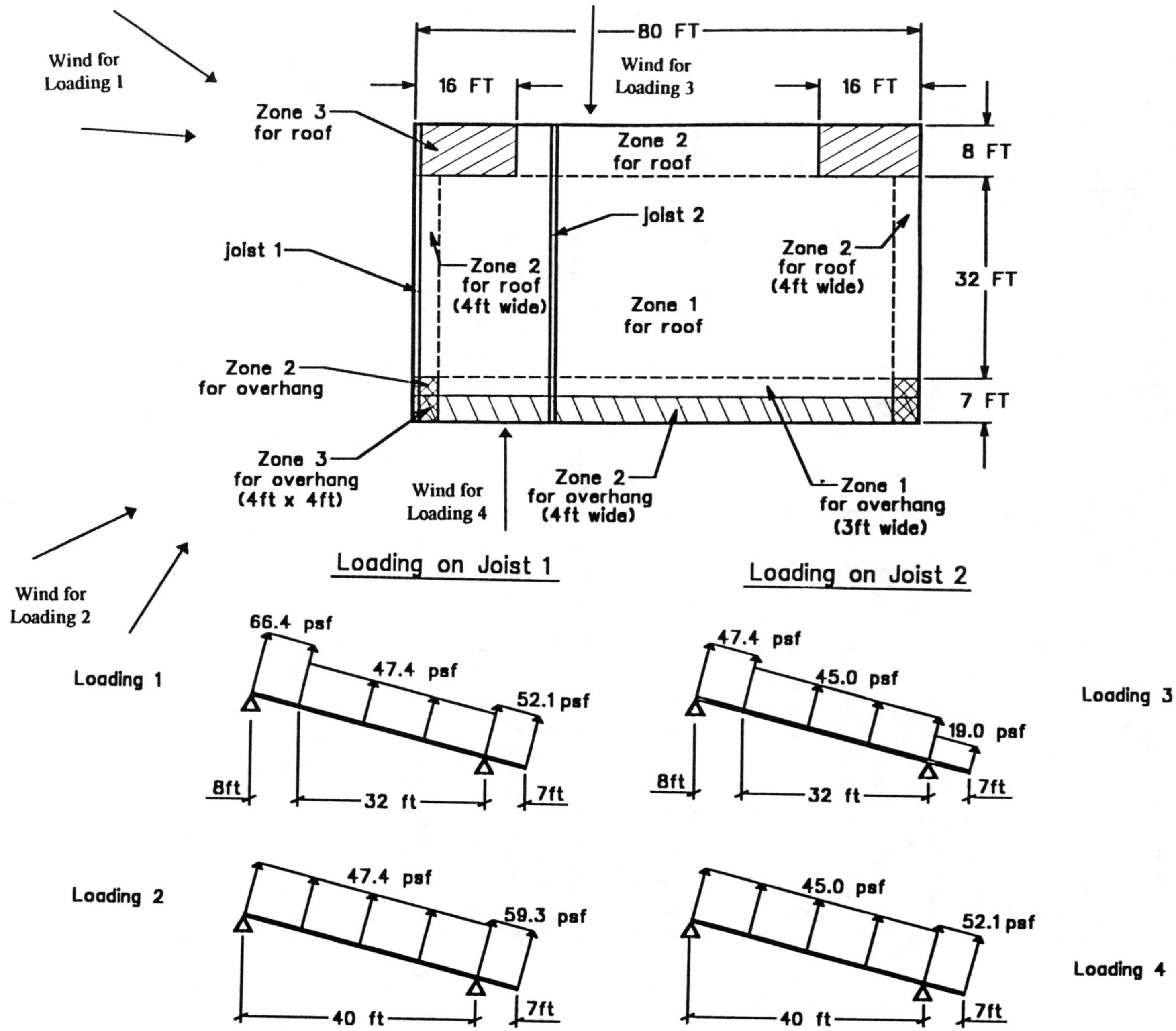

Figure 3.22 Design Pressures for Typical Joists and Pressure Zones for Roof Components and Cladding

3.6 EXAMPLE 6

In this example design wind forces for a tall billboard solid sign are determined. The example illustrates two items; (1) determination of G_f for a flexible structure, and (2) use of force coefficient for other structures. The billboard sign data are as follows:

Location: Interstate Highway in Iowa

Terrain: Flat and open terrain

Dimensions: 50 ft x 20 ft sign mounted on two 16 inch diameter steel pipe supports; bottom of the sign is 60 ft above ground.

Structural Characteristics: Tall flexible structure; estimated fundamental frequency is 0.7 Hz and critical damping ratio is 0.01.

(The natural frequency of a structure can be calculated in different ways. It has been predetermined for this example.)

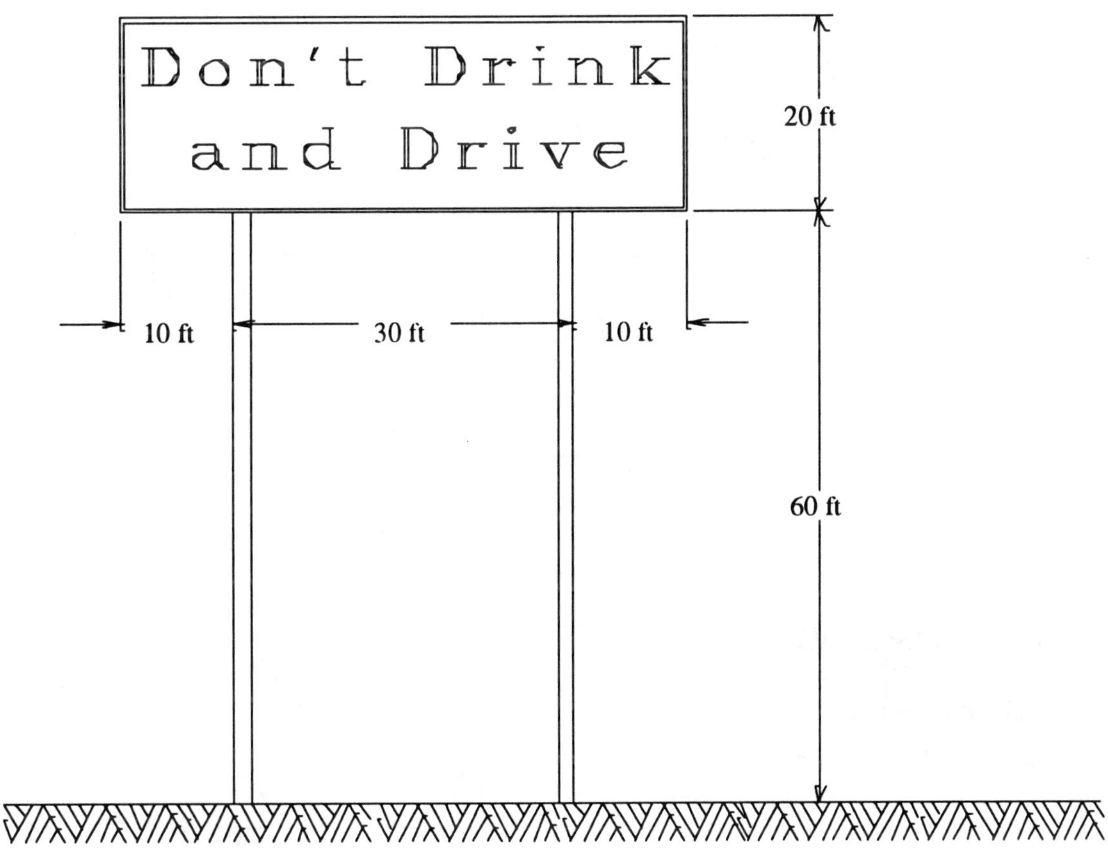

Figure 3.23 Dimensions of a Billboard Sign on an Interstate Highway

Exposure and Building Classification

The sign is in an open area, therefore use Exposure C.

Failure of sign represents low hazard to human life since it is located away from the highway and is not in populated area; the structure can be classified as Category I (Table 1-1).

Basic Wind Speed

The wind speed map in Figure 6-1 has only one value of wind speed in the middle of the country. Exact location of the sign in Iowa is not important. The basic wind speed V = 90 mph.

Velocity Pressures

The velocity pressures are computed using:

$$q_z = 0.00256 K_z K_{zt} V^2 I \text{ psf} \qquad \text{(Eq. 6-1)}$$

where V = 90 mph
I = 0.87 for Category I (Table 6-2)
K_{zt} = 1.0 because of flat terrain
K_z = values from Table 6-3 for z of 30, 60 and 80 ft. More divisions of z are not justified because loads on pipe supports are small compared to the ones on sign.

Velocity Pressures, psf

Height, ft	K_z	q_z, psf
30	0.98	17.7
60	1.13	20.4
80	1.21	21.8

Design Force for MWFRS

The equation given in Table 6-1 is as follows:

$$F = q_z G_f C_f A_f$$

where q_z is as determined previously
G_f is a gust effect factor to be calculated by a rational analysis (Section 6.6.2) because f < 1 Hz (Table 6-1). One rational analysis procedure is given in the commentary which will be used here.
C_f is force coefficient values from Tables 6-7 and 6-8
A_f = 20 x 50 = 1000 sq. ft; for normal and oblique wind see note 3 in Table 6-8.

Force Coefficient C_f

This sign qualifies as an above ground level sign with $\frac{M}{N} = 2.5$; $C_f = 1.2$. (Table 6-8)

The supports are round.
From Table 6-7:

$$D\sqrt{q_z} = 1.33\sqrt{17.7} = 5.6 > 2.5 \text{ and } \frac{h}{D} = \frac{60}{1.33} = 45$$

For moderately smooth surface, $C_f = 0.7$

Gust Effect Factor G_f

The gust effect factor G_f is determined from equation C6-9:

$$G_f = \frac{1 + 2gI_{\bar{z}}\sqrt{Q^2 + R^2}}{1 + 7I_{\bar{z}}}$$

where $I_{\bar{z}}$ is from Equation C6-6
g is taken as 3.5 (see page 159)
Q is determined from equation C6-7
R is determined from equations on page 159
$\bar{z}$ is the equivalent height of the structure, it is used to determine nominal value of $I_{\bar{z}}$; for buildings the recommended value is 0.6h, but for the sign it should be the middle of the billboard area or 70 ft.
c, l, α, ε, etc. are obtained from Table C6-6

$$I_{\bar{z}} = c\left(\frac{33}{\bar{z}}\right)^{1/6} = 0.2\left(\frac{33}{70}\right)^{1/6} = 0.176 \qquad \text{(Eq. C6-6)}$$

$$L_{\bar{z}} = l\left(\frac{\bar{z}}{33}\right)^{\varepsilon} = 500\left(\frac{70}{33}\right)^{1/5} = 581 \text{ ft} \qquad \text{(Eq. C6-8)}$$

$$Q^2 = \frac{1}{1 + 0.63\left[\frac{b+h}{L_{\bar{z}}}\right]^{0.63}} \qquad \text{(Eq. C6-7)}$$

$$= \frac{1}{1 + 0.63\left[\frac{50+20}{581}\right]^{0.63}} = 0.858$$

Note: In eq. C6 - 7, b and h are the dimensions of the sign.

$$\overline{V}_{\bar{z}} = \bar{b}\left(\frac{\bar{z}}{33}\right)^{\bar{\alpha}} \hat{V}_{ref} = 0.65\left(\frac{70}{33}\right)^{1/6.5}(90) = 65.67 \qquad \text{(Eq. on page 159)}$$

Note: $\hat{V}_{ref}$ is the basic wind speed

$$N_1 = \frac{n_1 L_{\bar{z}}}{\overline{V}_{\bar{z}}} = \frac{(0.7)(581)}{65.67} = 6.19$$

Note: n_1 is the fundamental frequency of the structure

$$R_n = \frac{7.465 N_1}{(1 + 10.302 N_1)^{5/3}} = 0.044 \qquad \text{(Eq. on page 159)}$$

$$\textit{For } R_h\text{ ; } \eta = \frac{4.6 n_1 h}{\overline{V}_{\bar{z}}} = \frac{(4.6)(0.7)(80)}{65.67} = 3.923 \qquad \text{(Eq. on page 159)}$$

Note: h is taken as 80 ft because resonance response depends on full height.

$$R_h = \frac{1}{\eta} - \frac{1}{2\eta^2}\left(1 - e^{-2\eta}\right) = 0.2224 \qquad \text{(Eq. on page 159)}$$

$$\textit{For } R_b\text{ ; } \eta = \frac{(4.6)(0.7)(50)}{65.67} = 2.452 \qquad \text{(assuming b = 50ft)}$$

$$R_b = 0.3253$$

$$\textit{For } R_d\text{ ; } \eta = \frac{(15.4)(0.7)(2)}{65.67} = 0.328 \qquad \text{(assuming depth d = 2 ft)}$$

$$R_d = 0.8130$$

$$R^2 = \frac{1}{\beta} R_n R_h R_b (0.53 + 0.47 R_d) \qquad \text{(Eq. on page 159)}$$

$$= \frac{1}{0.01}(0.044)(0.2224)(0.3253)\left[0.53 + (0.47)(0.8130)\right]$$

$$= 0.290$$

$$G_f = \frac{1 + 2(3.5)(0.176)\sqrt{0.858 + 0.290}}{1 + (7)(0.176)} \qquad \text{(Eq. C6-9)}$$

$$G_f = 1.039$$

Design Force

Force, $F = q_z(1.039)C_f A_f$

For each support:

for 0 to 30 ft $\quad F = 17.7(1.039)(0.7)(1.33) = 17.1$ plf

for 30 to 60 ft $\quad F = 20.4(1.039)(0.7)(1.33) = 19.7$ plf

for two supports, 0 to 30 ft: 34.2 plf; 30 to 60 ft: 39.4 plf

For the sign:

$$F = 21.8(1.039)(1.2)(50) = 1359 \text{ plf}$$

The force on the sign follows two cases;

1. force at geometric center
2. force at 0.2(50) = 10 ft on either side of geometric center (Note 3, Table 6-8)

Limitation

In certain circumstances for circular members, across-wind response due to vortex shedding can be critical. The Standard does not provide a procedure to assess across-wind response, but suggests obtaining guidance from recognized literature (Section 6.4.2.1).

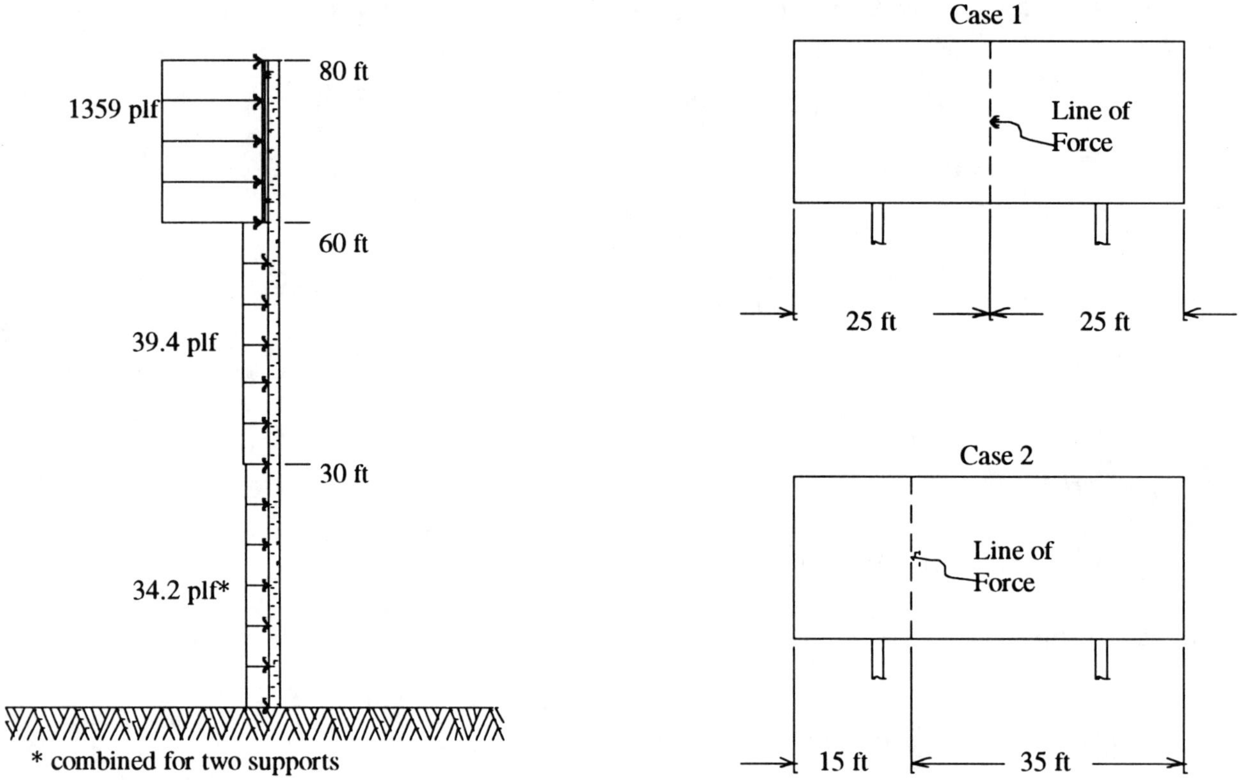

Figure 3.24 Design Forces for the Billboard Sign

<u>Force on Components and Cladding</u>

The equation given in Table 6-1 is

$$F = q_z G C_f A_f$$

The values of q_z are the same a MWFRS except the value of G = 0.85. The design forces can be determined using appropriate C_f and A_f for each component or cladding panel.

Chapter 4
BACKGROUND AND COMMENTARY

4.1 INTRODUCTION

This chapter addresses the use of the various tables and figures in the wind load provisions and provides additional information and background material not presented in the Standard. It is not the intention herein to address all items in detail; to do so would require a very lengthy document. For those users of the ***Guide*** who are interested in more detailed information on wind characteristics and wind loading effects, a comprehensive list of references is included at the end of this chapter.

4.2 DESIGN WIND LOADS

4.2.1 General

ASCE 7-95 recognizes two procedures for the determination of design wind loads for buildings and other structures as a whole or for individual components and cladding: 1) an analytical procedure, and 2) a wind tunnel procedure. Details of the analytical procedure are addressed in the following paragraphs.

4.2.2 Analytical Procedure

The determination of design wind pressures or forces for all cases covered by the Standard begins with the selection of the appropriate equation in Table 6-1. The calculations require three major elements as follows:

1) a velocity pressure, q_z or q_h
2) a gust effect factor, G or G_f
3) a pressure or force coefficient, C_p or C_f,
   or a combined coefficient, GC_p, GC_{pi} or GC_{pf}

Descriptions and the use of each of the above major elements are covered in subsequent sections of this chapter. In certain tables and figures of the Standard the gust effect factor and the pressure coefficient are given as a product or combined factor (GC_p, GC_{pi} and GC_{pf}). In these situations separate gust effect factors, G, are not to be determined or used. The reason for adopting this format is that modern wind tunnel tests are conducted using properly simulated turbulent boundary layers. Measured pressures and forces thus include both the steady and fluctuating components, and the peak values automatically include a gust effect factor. Prior to the introduction of the boundary layer wind tunnel in the 1960's, testing was carried out in aeronautical wind tunnels having smooth flow and a uniform velocity distribution. Pressure and force coefficients obtained from these earlier tests reflect only the steady component, and it is necessary to use a suitable gust effect factor to account for the fluctuating component. There are numerous applications for which only the older smooth-flow pressure and force coefficients are available.

The first part of Table 6-1 covers all buildings and other structures not classified as being flexible. Flexible buildings and other structures are defined as having a fundamental natural frequency less than 1 Hz. Also included in this classification are buildings and other structures having a height h exceeding four times the least horizontal dimension. The height-to-width criterion is somewhat arbitrary and is intended to save the designer the trouble of calculating the natural frequency in every case. If there is any reason to suspect that flexibility may be a problem, a gust effect factor, G_f, for MWFRS of flexible buildings should be determined as discussed later under **GUST EFFECT FACTORS**.

Table 6-1 also distinguishes between low-rise buildings (h ≤ 60 ft), buildings with h > 60 ft, and open buildings and other structures. An open building is defined as a structure having all walls at least 80% open. And finally, the design wind loading is obtained either for the main wind-force resisting system (MWFRS), or for components and cladding (C&C). The Standard requires that the wind load used in the design of MWFRS be not less than 10 lb/sq ft multiplied by the area of the building or structure projected on a plane normal to the wind direction. Similarly, the design force for open buildings and other structures shall be not less than 10 lb/sq ft multiplied by A_f, the area either normal to the surface or projected on a plane normal to the wind direction (the notes in Tables 6-6 through 6-10 indicate which area to use). Wind pressures obtained from Table 6-1 for the design of components and cladding are the net pressures resulting from the simultaneous application of the external and internal pressures. The Standard requires that the minimum design pressure for components and cladding be not less than 10 lb/sq ft acting either inward or outward. Taking the items in the order they are addressed in Table 6-1, we have the following procedures:

Nonflexible Buildings and Other Structures - MWFRS:

Low-rise buildings: The velocity pressure is q_h evaluated at mean roof height using Exposure C, regardless of the actual terrain roughness surrounding the site. Exposure categories are discussed later in this chapter. External pressure coefficients, GC_{pf}, are applied in accordance with the requirements of Fig. 6-4, the appropriate internal pressure coefficients, GC_{pi}, are selected from Table 6-4, and the design pressures, p, are calculated from the formula given in Table 6-1. Note that if the building is actually sited in Exposure B for all wind directions, the design wind pressures can be reduced by multiplying by the factor 0.85. The reason for using a velocity pressure q_h based on Exposure C is that the external pressure coefficients, GC_{pf}, in Fig. 6-4 were obtained from wind tunnel tests in which the boundary layer was modeled consistent with Exposure C (Davenport et al., Phase I, II & III, 1977, 1978). Additional studies have shown that low-rise buildings sited in Exposure B experience some direct shielding from surrounding construction, and this is the reason for the reduction factor of 0.85 (Ho 1992).

The internal pressure is included in the determination of the design wind loading for MWFRS since it generally is a major part of the loading on main structural assemblies such as frames in low-rise buildings. When determining base shear the internal pressures automatically cancel, and their net effect is zero. However, this is not the case when determining uplift forces. Application of the two loading cases given in Fig. 6-4 for design of the MWFRS is covered later in a discussion of Fig. 6-4.

Buildings of all heights: The velocity pressure is q_z for the windward wall at height z above ground and q_h at mean roof height for the leeward wall, the side walls, the roof, and the internal pressure. Accordingly, the internal pressures automatically cancel when determining base shear as described for low-rise buildings. Depending on the characteristics of the MWFRS used, the internal pressure may or may not be important. Generally, the internal pressure is an important part of the wind loading for roof framing. The velocity pressures are determined on the basis of the actual wind exposure for any given wind direction, either A, B, C, or D. The choice of number of increments of z to use in calculating wind loads on the windward wall is left to the designer. For any given increment of height, the pressure is calculated at the top of the increment. Obviously, the accuracy of the load estimate will increase with the number of increments used. Appropriate values of C_p are obtained from Fig. 6-3, and GC_{pi} is obtained from Table 6-4. As will be discussed later under **GUST EFFECT FACTORS**, the value of G is 0.8 for Exposures A or B and 0.85 for Exposures C or D.

Open buildings and other structures: The design wind force, F, acts on the area A_f. The specific definition of A_f is included in the table from which the appropriate force coefficients, C_f, are obtained. Again, the value of G is 0.8 for Exposures A or B, and 0.85 for Exposures C or D.

Nonflexible Buildings and Other Structures - Components & Cladding:

Low-rise buildings with $h \leq 60$ ft: The determination of design pressures for C&C proceeds in exactly the same manner as with pressures for the design of MWFRS using q_h at the mean roof height. As with the MWFRS, Exposure C is assumed even though the actual exposure may fit some other category. Values of GC_p are dependent upon the effective wind area and are obtained from Figs. 6-5, 6-6 and 6-7. The effective wind area for components and cladding panels is the span length multiplied by an effective width that need not be less than one-third the span length. The reason for this limitation on minimum effective width is that the intensity of area-averaged pressure fluctuations decreases with increasing area over which the pressures are averaged. The one-third rule is intended to exclude narrow rectangles of small area that would tend to overestimate GC_p. Typically, the one-third rule is applied to the determination of GC_p values for the design of closely-spaced roof trusses or wall studs. For cladding fasteners, the effective wind area is the area secured by an individual fastener. Figs. 6-5, 6-6 and 6-7 indicate maximum values of GC_p for effective wind areas of 10 sq ft or less and minimum values of GC_p for effective areas larger than 100 to 500 sq ft.

In recognition of the fact that area-averaged wind pressures decrease with increasing area, the Standard allows major structural components supporting tributary areas (as opposed to effective wind areas) greater than 700 sq ft to be designed using the provisions for MWFRS. If the building is actually sited in Exposure B for all wind directions, the design wind pressures obtained for Exposure C may be multiplied by the factor 0.85.

Buildings with $h > 60$ ft: The procedure for calculating design wind pressures for components and cladding is similar to that for MWFRS with the exception that the external and internal pressure coefficients are summed prior to multiplication by the velocity pressure, q_z or q_h. The primary reason for this is convenience for the user of the Standard. For positive external pressures, q_z is used for the velocity pressure, and for negative external pressures it is q_h. Values of q_h are based on the most critical exposure category representative of the building site with the exception that Exposure B shall be used for buildings and other structures sited in terrain representative of Exposure A. The reason for this limitation is the uncertainty of channeling caused by tall buildings and its local effects on the relatively smaller velocity pressures for Exposure A. Values of the pressure coefficients, GC_p, are obtained from Fig. 6-8, and as with design of the MWFRS, values of the internal pressure coefficients, GC_{pi}, are obtained from Table 6-4. Again, major structural components supporting tributary areas greater than 700 sq ft may be designed using the provisions for MWFRS.

The external pressure coefficients for roofs in Fig. 6-8 are for roofs with $\theta \leq 10$ degrees. For other roof angles and geometry, it is permissible to use values of GC_p from Fig. 6-5B and the attendant q_h based on Exposure C. However, multiplication by the factor 0.85 for Exposure B is not allowed. In designing components and cladding for buildings having a mean roof height greater than 60 ft but less than 90 ft, values of GC_p from Figs. 6-5, 6-6, and 6-7 may be used only if the height-to-width ratio of the building is ≤ 1, q is taken as q_h, and Exposure C is used for all terrain.

Open buildings and other structures: The determination of design wind pressures for components and cladding proceeds in exactly the same way as for MWFRS, using the velocity pressures, force coefficients and gust factors. Exposure B shall be used for the design of C&C for open buildings or other structures sited in Exposure A. Again, the 700 sq ft rule applies to the design of major structural components supporting tributary areas of this size or larger.

Flexible Buildings and Other Structures - MWFRS:

Buildings: The determination of design wind loads for the MWFRS in flexible buildings employs the same procedure as used for nonflexible buildings with the exception that the gust effect factor, G_f, must be calculated using a rational analysis that incorporates the dynamic properties of the MWFRS. One such method of rational analysis is included in the Commentary, and this is discussed later under **GUST EFFECT FACTORS**. Also, it is not necessary to consider internal pressures as the along-wind response of the overall MWFRS is the issue. The resulting wind loads, when applied to the building as static loads, should produce the same load effects as the actual dynamic loads.

Other structures: The procedure is the same as for nonflexible open buildings and other structures, except that the gust effect factor, G_f, must be calculated using a rational analysis that incorporates the dynamic properties of the MWFRS.

Flexible Buildings and Other Structures - Components & Cladding:

Buildings: The procedure is identical to that used for components and cladding for nonflexible buildings. However, the Standard does not allow the optional use of GC_p values from Figs. 6-5, 6-6, and 6-7 when the mean roof height is greater than 60 ft, but less than 90 ft. Again, the 700 sq ft rule for tributary area applies to the design of major structural components.

Other structures: The procedure is identical to that for other structures that are nonflexible, including the choice of gust effect factor, G, the force coefficients, C_f, and the 700 sq ft rule.

4.3 VELOCITY PRESSURE

The velocity pressure, q_z, at height z above ground level is fundamental to the determination of design wind pressures and forces and is calculated from Eq. 6-1 as follows:

$$q_z = 0.00256 K_z K_{zt} V^2 I \text{ (lb/sq ft)} \qquad \text{(Eq. 6-1)}$$

The numerical constant 0.00256 reflects the mass density of air for the standard atmosphere and the dimensions associated with velocity pressure, q_z, in pounds per square foot and wind speed, V, in miles per hour as indicated in Eq. C2. The mass density of air varies with altitude, air temperature and humidity, and Table C6-1 lists a range of values (minimum, average, maximum) for altitudes up to 10,000 ft. If the numerical constant in Eq. 6-1 is to be changed on the basis of higher altitude, the maximum ambient air density at that altitude should be used to account for variations in temperature and humidity. Descriptions of the velocity pressure exposure coefficient K_z, the topographic factor K_{zt}, the basic wind speed V, and the importance factor, I, are given in the following subsections. The topographic factor represents a wider scope of application than was possible with previous editions of the Standard, and the form of the importance factor, I, has been changed from ASCE 7-93. Previously the importance factor, I, was applied to the velocity in the form $(VI)^2$ rather than to the velocity pressure as given in equation 6-1.

4.3.1 Basic Wind Speed

The basic wind speed, V, used in the determination of design wind loads is given in Fig. 6-1. The values are 3-second gust speeds in miles per hour at 33 ft above ground for a category C wind exposure. The speeds are associated with a probability of 0.02 of being equaled or exceeded in any given year (50-yr mean recurrence interval). Adoption of the 3-second gust speed in lieu of the traditional fastest-mile wind speed was prompted by the decision of the National Weather Service to abandon the fastest-mile as a standard measure of daily maximum wind speed. This change in basic wind speed led to a comprehensive analysis of peak gust data and the correlation of gust speeds with the traditional fastest-mile speeds. While the adoption of the 3-second gust speed has caused certain coefficients to change significantly from the values given in past versions of the Standard, this change in definition of the basic wind speed has not resulted in any systematic increase or decrease in the design wind loads.

The wind speed map of Fig. 6-1 is based on peak gust data collected at 485 weather stations where at least 5 years of data were available. Also used in the preparation of the wind speed map were predicted hurricane speeds along the U.S. Gulf and Atlantic Coasts. For nonhurricane (extratropical) regions, gust data were assembled from several stations in state-size areas, and the resulting series of extremes were then fit using a Fisher-Tippett Type I extreme value distribution (Peterka 1992, Peterka and Shahid 1993). The variation in the resulting 50-year wind speeds across the eastern 3/4 of the contiguous 48 states was insufficient to justify wind speed contours, hence the large 90 mph region shown on the map. In addition, the division between the 90 and 85 mph regions, which follows state boundaries on the map of Fig. 6-1, was sufficiently close to the 85 mph contour that there was no statistical basis for choosing some other boundary between the two wind speed regions.

Extreme wind speeds in hurricane-prone regions such as the Atlantic and Gulf Coasts present a special problem. Because hurricanes make landfall infrequently and affect areas of relatively limited extent, the number of reliable wind speed measurements at any given site is insufficient to support a valid statistical analysis. To circumvent this problem, models relating wind speed, radial distance, and pressure gradient are used with Monte Carlo simulations whose statistics match the historical records of hurricane occurrences, central pressure and storm tracks (Batts et al. 1980, Georgiou et al. 1983, Vickery and Twisdale 1993). The wind speeds resulting from a sufficiently large number of these simulations form the basis for the wind speeds indicated in Fig. 6-1 for the Atlantic and Gulf Coasts. In converting the results of these simulations to 3-second gust speeds, use was made of the gust factors for hurricanes indicated by the dotted (upper) curve in Fig. C6-1 (Krayer and Marshall 1992). This curve is based on a number of wind speed records obtained in actual hurricanes that suggest the gust factors are slightly higher than those associated with extratropical storms (Durst 1960). The hurricane wind speed contours plotted off the Atlantic Coast in Fig. 6-1 are placed there only for the purpose of interpolation and are based on Exposure C terrain.

The wind speeds specified in the Standard do not include tornadic winds. This is because the expected wind speeds in tornadoes for most regions of the United States are associated with annual probabilities that are much smaller than 0.02, the wind speed probability generally assumed in the design of ordinary buildings and other structures and the probability associated with the wind speeds given in Fig. 6-1. To put this another way, the tornadic wind speeds associated with an annual probability of 0.02 of striking a point target are substantially less than the wind speeds given in Fig. 6-1 (Murray and McDonald 1993, Marshall and Yokel 1995). Extensive documentation of building damage caused by tornadoes and the analysis of collected data indicate that in approximately half of the recorded tornadoes the speeds are less than the basic wind speeds of Fig. 6-1. Consequently, if a building or other structure is designed for the wind loads specified by the standard, it is likely to resist approximately one-half of all tornadoes without significant structural damage. Tornadic wind speeds associated with an annual probability of $1x10^{-5}$ (mean recurrence interval of 100,000 years) are shown in Fig. C6-1A. Sufficient information is available for development of design criteria and actual design of tornado-resistant shelters and buildings that house facilities essential for post-disaster operations and recovery (Wen and Chu 1973, Mehta et al. 1976, Minor et al. 1977, Minor 1982, McDonald 1983).

Although the wind speeds given in Fig. 6-1 are valid for most regions of the country, they may not apply to mountainous terrain, river gorges, ocean promontories or to the special wind regions indicated on the map of Fig. 6-1 where local effects may cause substantially higher speeds than those shown on the map. Typical of these special wind regions is the Front Range of the Rocky Mountains in Colorado where the mountain lee-wave can bring high-altitude, fast-moving air to ground level near the base of the mountains. For these special situations, or whenever there is reason to believe the wind speeds indicated in Fig. 6-1 do not reflect adequately the local wind climate, the user of the Standard should review the regional climatic data and seek expert advice from a consulting meteorologist or wind engineer.

In conducting an assessment of wind speed records, it is important that recognized statistical analysis procedures for extreme values be followed carefully. When using local data, sampling errors (errors associated with the limited years of annual extremes) can lead to large uncertainties in the 50-year wind speed, and care and conservatism should be exercised. The effects of certain isolated topographic features may be handled by the topographic factor, K_{zt}, which is described later. Wind tunnel tests may be warranted when there is a need to assess the effects of complex terrain or other unusual features.

4.3.2 Velocity Pressure Exposure Coefficients K_h and K_z

The velocity pressure exposure coefficient, K_z, in Eq. 6-1 accounts for the variation of wind speed with height above ground and with ground roughness. K_h is a special case of the velocity pressure exposure coefficient K_z, namely the coefficient evaluated at the mean roof height of a building or at the height of a structure as called for in Table 6-1. Values of K_h and K_z are listed in Table 6-3 for heights up to 500 ft and for four categories of ground roughness, designated as Exposures A, B, C, and D.

Values of the velocity pressure exposure coefficient decrease with decreasing height, reaching a theoretical value of zero at the ground surface. However, because of the flow complexities due to high turbulence near the ground, K_z is assumed to be constant for $z \leq 15$ ft. Table 6-3 lists values of the velocity pressure exposure coefficient for values of z up to 500 ft. For heights greater than 500 ft, Eq. C3a in the Commentary may be used, and values of the exposure category constants α and z_g can be obtained from Table C6-2. Eq. C3a is valid up to the gradient height, z_g, and K_z is constant above this height.

Users of ASCE 7-93 will notice significant changes in the values of K_h and K_z listed in Table 6-3 of ASCE 7-95. This is due to the fact that the distribution of the 3-second gust speed with height above ground and with terrain roughness differs from that of the fastest-mile wind speed which was the basis for wind pressure and load calculations in previous editions of the Standard. As can be seen from Table C6-2, values of the gradient height, z_g, the height above which the ground roughness has no significant effect on the mean wind speed, have been retained. However, the values of α used in the power-law representation of the mean velocity profiles have been revised. For a detailed discussion of changes to the velocity pressure exposure coefficient, K_z, see the Commentary to ASCE 7-95.

4.3.2.1 Exposure Categories

As with previous versions of the Standard, four separate categories of wind exposure are recognized. In selecting the category that best describes the surface roughness upwind of the site in question, account shall be taken of variations in roughness that arise from natural topography and vegetation as well as from any constructed features. As can be seen from Table 6-3, the choice of wind exposure category can have a significant impact on the velocity pressure exposure coefficient, K_z, and likewise on the velocity pressure q_z. To aid the user of the Standard in this selection process, which is necessarily subjective, the following guidelines are provided:

Exposure C

The Standard describes this exposure as open terrain with scattered obstructions having heights generally less than 30 ft. This category includes flat open country and grassland. Exposure category C is assumed for the basic wind speeds given in Fig. 6-1, and the Standard requires the use of this exposure with the pressure coefficients GC_{pf} in Fig. 6-4, GC_{pi} in Table 6-4, and GC_p in Figs. 6-5 to 6-7. The reason for this requirement is that these pressure coefficients are based on wind tunnel studies conducted in boundary-layer flows consistent with Exposure C. Also, this exposure category should be used whenever the actual terrain does not fit any of the other exposure category descriptions.

Exposure A

This exposure category includes large city centers with at least 50% of the buildings having a height in excess of 70 ft. Use of this exposure is limited to those areas for which terrain representative of Exposure A prevails in the upwind direction for a distance of at least 1/2 mile or 10 times the height of the building or other structure, whichever is greater. Exposure A can be applied only to the design wind loads for MWFRS. Exposure A cannot be applied to the determination of wind pressures for the design of components and cladding. This restriction is due to the fact that the effects of channeling or increased buffeting due to the building or structure being located near the wake of adjacent buildings can more than offset the difference in velocity pressure between Exposures A and B. Also, care should be taken to ensure that Exposure A applies for all wind directions. It is possible that heavily built-up areas in one direction could qualify for Exposure A while other directions would best be described by one or more of the other exposure categories.

Exposure B

The Standard describes this exposure as urban and suburban areas, wooded areas, or other terrain with numerous closely spaced obstructions having the size of single-family dwellings or larger. Use of this exposure category is to be limited to those areas for which terrain representative of Exposure B prevails in the upwind direction for a distance of at least 1,500 ft or 10 times the height of the building or other structure, whichever is greater. When low-rise buildings with mean roof height h ≤ 60 ft are sited in Exposure B for all wind directions, the wind loads for MWFRS using Fig. 6-4 and the wind loads for components and cladding calculated on the basis of Exposure C can be multiplied by the factor 0.85. Numerous post-storm damage investigations have shown that open areas such as athletic fields directly upwind of buildings otherwise sited in Exposure B can cause damaging increases in wind loading. Therefore, care should be exercised when selecting Exposure B to ensure that it does in fact apply for all wind directions.

Exposure D

This exposure category includes unobstructed areas exposed to wind flowing over open water for a distance of at least 1 mile. This exposure applies only to those buildings and other structures exposed to the wind coming from over the water, and the exposure extends inland from the shoreline a distance of 1,500 ft or 10 times the height of the building or structure, whichever is greater. Buildings and other structures located away from beaches and having structures between them and the water need not be designed on the basis of Exposure D. Additionally, Exposure D does not apply to the design of components and cladding for buildings with mean roof height h ≤ 60 ft as the Standard requires the use of Exposure C for the design of such elements.

4.3.3 Topographic Factor K_{zt}

When the wind in the lower levels of the boundary layer encounters a topographic feature such as a ridge, escarpment or hill, blockage effects cause the flow nearest the feature to accelerate with maximum speed-up occurring at the crest. The flow at points sufficiently far from the topographic feature, either laterally or vertically, is unaffected. Post-storm damage surveys have shown a strong correlation of damage intensity with these speed-up effects. All other conditions being equal, speed-up effects are most significant for low-rise buildings. The Standard addresses this speed-up phenomenon by means of a topographic factor, K_{zt}, in the formula for velocity pressure (Eq. 6-1). In the absence of a hill or similar topographic feature, K_{zt} has a value of unity and may be neglected when calculating q_z. The procedure is limited to isolated 2-dimensional ridges, escarpments and axisymmetric hills in Exposure B, C, or D where the upwind terrain is free of such topographic features for a distance of 50 times the height of the feature or one mile, whichever is smaller, as measured from the crest of the feature.

In addition to the generic shape of the topographic feature, the parameters H and L_h are required. H is the height of the feature relative to the upwind terrain, and L_h is the horizontal distance upwind of the crest to where the difference in ground elevation is half the height of the topographic feature. This definition of L_h simplifies the description of those features having a varying windward slope or a beginning (toe) of slope that is poorly defined. The calculation of K_{zt} involves three multipliers, K_1, K_2 and K_3, and the formula

$$K_{zt} = [\ 1 + K_1K_2K_3\ ]^2$$

The multipliers are determined by interpolation from Fig. 6-2 and are always equal to or greater than zero. For a given topographic feature, K_1 depends upon the ratio H/L_h which is twice the value of the average upwind slope of the hill or escarpment. K_2 depends upon the ratio x/L_h where x is the horizontal distance from the crest to the building site, either upwind or downwind of the crest. K_3 depends upon the ratio z/L_h where z is the height above the local ground level at the building site. When H/L_h is greater than 0.5 (an average windward slope or gradient of 0.25), the multipliers are no longer dependent upon the horizontal dimension L_h. When this happens, 2H is substituted for L_h in the ratios H/L_h, x/L_h, and z/L_h.

To reduce the demands placed on the user of the Standard, speed-up effects need not be considered when H/L_h is less than 0.2, when H is less than 15 ft for Exposure D, when H is less than 30 ft for Exposure C, or when H less than 60 ft in Exposure B. Speed-up effects need not be considered for Exposure A. In addition, speed-up effects need be considered only for buildings and other structures located on the upper half of isolated ridges, escarpments and hills. The provisions for speed-up assume the direction of the wind is along the line of maximum slope. Although procedures are available for other relative wind directions, the calculations are more complicated. More information on speed-up effects may be found in the references (Jackson and Hunt 1975, Walmsley et al. 1986, Lemelin et al. 1988).

4.3.4 Importance Factor

The importance factor, I, is used to adjust the design wind load to a level consistent with the building classifications described in Table 1-1 of the Standard. The values of I for wind loading in Table 6-2 range from 0.87 for Category I structures that represent a low hazard to human life in the event of failure, to 1.15 for Category III and IV structures that represent a substantial hazard to human life in the event of failure or must serve some essential function such as a hospital or a fire station. All other buildings and structures are designated as Category II structures with I = 1.00. In ascending order, these importance factors correspond to annual probabilities of 0.04, 0.02 and 0.01 of the design wind speed being equaled or exceeded. The corresponding mean recurrence intervals are 25, 50 and 100 years.

In previous editions of the Standard the importance factor was applied to the basic wind speed in the form $(VI)^2$ rather than to the velocity pressure as indicated by Eq. 6-1. Therefore, use of the importance factor is equivalent to multiplying the basic wind speed by the square root of I, and the corresponding conversion factors are 0.93, 1.00 and 1.07. Referring to Table C6-5 of the Commentary, it can be seen that these same conversion factors appear in the column headed "V = 85 to 100 mph," and the corresponding mean recurrence intervals are 25, 50 and 100 years.

Because the probability distributions for extreme wind speeds in hurricanes and in extratropical storms are different, a hurricane importance factor is required to ensure the same probability of overload in hurricane-prone regions as in other regions. To accomplish this, ASCE 7-93 specified a factor of 1.05 to be applied to the basic wind speed at the hurricane oceanline, with the factor decreasing linearly to a value of 1.0 at a distance of 100 miles inland. This same factor has been retained in ASCE 7-95 and is incorporated in the wind speeds of Fig. 6-1 for the hurricane-prone Gulf and Atlantic Coasts. The factor decreases linearly to a value of 1.0 at the 90 mph contour in Fig. 6-1. Therefore, it is possible to adjust the velocity pressure, q_z, based on the 50-yr wind speeds of Fig. 6-1, to be consistent with annual probabilities of 0.04 or 0.01 (25- and 100-yr MRI) by simply multiplying the velocity pressure from Eq. 6-1 by the importance factors 0.87 or 1.15, respectively, as given in Table 6-2. This approach neglects some minor differences in the wind speed distributions for Alaska and for the extratropical region of the contiguous 48 states.

For applications that require wind speeds associated with mean recurrence intervals other than 50 years, the 50-year speeds of Fig. 6-1 are multiplied by the appropriate conversion factors listed in Table C6-5 in the Commentary. Because the 50-year wind speeds for the hurricane-prone regions of Fig. 6-1 have been multiplied by the factor 1.05, these same wind speeds should be divided by 1.05 (or the appropriate linearly reduced value) prior to adjusting them for other mean recurrence intervals. The probability P_n that the wind speed associated with a certain annual probability P_a will be equaled or exceeded at least once during an exposure period of n years is given by Eq. C4, and values of P_n for various values of P_a and n are given in Table C6-4 of the Commentary. Note that with proper choice of load factors or allowable stresses for design, the probability of failure is substantially less than the probability of exceeding the design wind speed.

4.4 GUST EFFECT FACTORS

The effects of velocity fluctuations (wind gusts) on the along-wind loading of buildings and other structures and correlation of gust size with buildings and other structure size are accounted for by the use of a gust effect factor, G. This factor, when denoted as G_f, also accounts for along-wind loading effects due to dynamic amplification for flexible buildings and structures. It does not include allowances for across-wind loading effects, vortex shedding, instability due to galloping or flutter, or dynamic torsional effects. For structures susceptible to loading effects that are not accounted for in the gust effect factor, information should be obtained from recognized literature (Kareem 1985, 1992, Solari 1993, Kareem and Smith 1993).

As was noted earlier in describing the Analytical Procedure, it is current practice to conduct wind tunnel studies in which both the boundary layer and the building or other structure under test are correctly modeled. In such cases the pressure data contain both the steady and fluctuating components, and the resulting pressure coefficients are designated by GC_p, indicating that gust effects are included. Wherever the Standard calls for the use of GC_p (Figs. 6-5 to 6-8), GC_{pf} (Fig. 6-4) or GC_{pi} (Table 6-4), a separate gust effect factor, G, is not to be used.

Prior to the introduction of the modern boundary-layer wind tunnel, it was common practice to test building models in a uniform air stream of low turbulence and to account for the turbulence in the natural wind by means of a separate gust effect factor G. This is the approach to be followed when using the pressure and force coefficients of Fig. 6-3 and Tables 6-5 to 6-10. In ASCE 7-93, these factors were termed "gust response factors," and their values ranged from 1.00 to 2.36. Consistent with the variation of turbulence in the boundary layer, these gust response factors decrease with increasing height above ground and increase with increasing terrain roughness. Adoption of the 3-second gust as the basic wind speed in ASCE 7-95 has resulted in a new set of gust effect factors whose values differ significantly from the old gust response factors of ASCE 7-93. As is described in the Commentary of the Standard, there are three categories of gust effect factors: (1) Rigid structures--simplified method; (2) Rigid structures--complete analysis; and (3) Flexible or dynamically sensitive structures. A description of each category follows.

Category I: Rigid Structures--Simplified Method

This category of gust effect factor applies to the design of MWFRS for buildings of ordinary size and having a fundamental natural frequency equal to or greater than 1 Hz. The general expression for the gust effect factor, G, is given in the Commentary by Eq. C6-5. The dimensions of the building are assumed to be of the order of the integral length scale of turbulence or smaller, resulting in G = 0.8 for Exposures A and B, and G = 0.85 for Exposures C and D. The reason for a gust effect factor, G, of less than 1 is that the use of the 3-second gust as the basic wind speed overcompensates for the area-averaged gust pressures acting on typical exterior surfaces. The fact that G is smaller for Exposures A and B than it is for Exposures C and D reflects the increased turbulence at higher frequencies for the rougher exposures, resulting in smaller integral length scales (average size of gusts) and reduced effectiveness of gusts in causing along-wind response.

Following the adoption of the 3-second gust as the basic wind speed and the corresponding adjustment of the velocity pressure exposure coefficients in Table 6-3, a calibration of the new wind load provisions against those of ASCE 7-93 was carried out. This calibration involved the comparison of calculated design wind loads for MWFRS in two different rigid buildings in 25 cities across the United States, using the requirements of Fig. 6-3. From this calibration it was found that the new wind load provisions displayed a bias toward higher along-wind forces on the MWFRS. To correct for this bias, a factor of 0.9 was applied to the gust effect factor as can be seen in Eq. C6-5. This factor is included in the values of G = 0.8 and G = 0.85 noted earlier.

Category II: Rigid Structures—Complete Analysis

Buildings and other structures to which this category of gust effect factor applies are assumed to be large in comparison to the integral length scale of turbulence. This leads to a reduction in the background response as can be seen from Eq. C6-7 in the Commentary, and this reduces the magnitude of the gust effect factor given by Eq. C6-5. Therefore, reductions in the along-wind loading of the MWFRS can be realized for buildings and other structures whose overall dimensions are large in comparison to the integral length scale of turbulence. However, it is important to understand that this category of gust effect factor only applies to buildings and other structures for which the fundamental natural frequency is greater than 1 Hz.

Category III: Flexible or Dynamically Sensitive Structures

For flexible buildings and other structures, the gust effect factor is designated as G_f. Values of G_f may be calculated by any rational procedure, one such procedure being included in the Commentary of the Standard, beginning with Eq. C6-9. From the definition and procedure outlined, it is possible to predict the along-wind response of a flexible structure, e.g., maximum displacement, rms acceleration, and peak acceleration.

Across-wind response is a result of vortex shedding or instability due to galloping or flutter. The Standard does not provide a procedure to assess across-wind response. Limitations of analytical procedure in Section 6.4.2.1 suggest that the designer should refer to recognized literature pertaining to across-winds effects or use wind tunnel procedure. Advice from a wind engineer should be obtained to assess severity of the across-wind problem.

4.5 PRESSURE AND FORCE COEFFICIENTS

4.5.1 General

The following sections describe the various pressure and force coefficients given in the Standard and certain conditions and limitations regarding their use. However, these descriptions are by no means comprehensive and strict attention to the footnotes in the tables and figures is advised.

4.5.2 External Pressure Coefficients for MWFRS, Enclosed or Partially Enclosed Buildings of All Heights

Pressure coefficients for design wind loads on MWFRS of enclosed or partially enclosed buildings of all heights are given in Fig. 6-3. The coefficients, C_p, were obtained in uniform flows of low turbulence and, therefore, are to be used in conjunction with a gust effect factor, G, which is 0.80 for Exposures A and B and 0.85 for Exposures C and D. For flexible buildings an appropriate G_f is to be determined by rational analysis. The pressure coefficients were determined for two orthogonal wind directions aligned with the principal axes of the building (normal and parallel to the ridge line). Thus, the resulting wind loads are directional loads as opposed to envelope loads. Note that a reduction in roof load is allowed for flat roof areas ($\theta < 10^0$) when h/L is equal to or greater than 1.0. Also note that two values of C_p are possible on the windward slope of roofs with certain roof angles. This results from the fact that the wind flow over the windward half of the roof may either separate near the eave or may remain fully attached. Both possibilities should be checked.

4.5.3 External Pressure Coefficients for MWFRS, Enclosed or Partially Enclosed Buildings With Mean Roof Height Less Than or Equal to 60 ft

The external pressure coefficients, GC_{pf}, given in Fig. 6-4 may be used for the determination of wind loads on MWFRS for enclosed or partially enclosed low-rise buildings with mean roof height h less than or equal to 60 ft. When the roof angle θ is less than or equal to 10°, h shall be set equal to the eave height. To obtain the critical wind actions to be resisted by the MWFRS, two separate loading conditions, Case A and Case B, must be considered with the building rotated in 90° increments so that each corner in turn becomes the windward corner while the loading patterns shown in Fig. 6-4 remain fixed. The procedure to be followed is illustrated in Fig. C6-2 of the Commentary.

The values of GC_{pf} given in Fig. 6-4 are such that when applied to the building as directed, the resulting structural actions (base shear, overturning, uplift, bending and torsion) will envelope those actions determined from building models tested in a wind tunnel with a properly modeled Exposure C boundary-layer flow. For this reason, certain values of GC_{pf} in Fig. 6-4 may appear to be inconsistent with values of GC_p presented elsewhere in the Standard. Even though it is not stated explicitly in Fig. 6-4, the pressure coefficients, GC_{pf}, are associated with q_h based on Exposure C, irrespective of the actual terrain surrounding the building site. For buildings sited within Exposure B for all wind directions, the calculated pressures may be multiplied by 0.85.

Although they are not directly comparable, the values of GC_{pf} listed in Fig. 6-4 are consistent with the original coefficients obtained from wind tunnel tests (Davenport et al., Phase I, II & III, 1977, 1978). Certain adjustments were required to account for differences in the definition of basic wind speed and the design format. The National Building Code of Canada (NRCC,1995) coefficients are based on the hourly mean speed, and include a factor of 0.8 to account for wind directionality among other things. In adjusting the coefficients to the ASCE 7-95 format, the ratio of the 3-second gust speed to the hourly mean speed was assumed to be 1.53, and the 0.8 factor was removed. Thus the values of GC_{pf} in Fig. 6-4 are equal to the coefficients in the National Building Code of Canada multiplied by the factor $(1/1.53)^2/(0.8) = 0.53$.

The original coefficients were intended for use with framing for traditional pre-engineered buildings with single-story moment-resisting frames in one of the principal directions and diagonal bracing in the other principal direction. Application of the coefficients was later extended to single-story moment-resisting frames with interior columns (Kavanagh et al. 1983). Recent wind tunnel tests have demonstrated that the pressure coefficients in Fig. 6-4 can be applied to low-rise buildings with structural systems consisting of shear walls and roof trusses (Isyumov and Case 1995).

4.5.4 External Pressure Coefficients for Loads on Components and Cladding, Enclosed or Partially Enclosed Buildings With Mean Roof Height Less Than or Equal to 60 ft

Figs. 6-5, 6-6 and 6-7 present values of external pressure coefficients, GC_p, for wind loads on components and cladding for enclosed or partially enclosed buildings with mean roof height less than or equal to 60 ft. The values of GC_p in these figures represent the maximum or envelope values acting on a range of effective wind areas and are based on both wind tunnel and full-scale measurements (Eaton and Mayne 1975, Best and Holmes 1978, Stathopoulos 1981, Stathopoulos and Mohammadian 1986, Surry and Stathopoulos 1988, Stathopoulos and Saathoff 1991, Saathoff and Stathopoulos 1992, Stathopoulos and Luchian 1992). In each case the pressure coefficients are to be used with the velocity pressure, q_h, based on Exposure C even though the actual exposure may match some other category. The wind tunnel tests show that certain building features lead to reduced loads. When calculating loads on walls, values of GC_p are to be reduced by 10% when the roof angle θ is less than or equal to 10°. If a parapet equal to or higher than 3 ft is provided around the perimeter of the roof with θ less than or equal to 10°, roof Zone 3 shall be treated as roof Zone 2. Also, values of GC_p for roof overhangs include pressure contributions from both the upper and lower surfaces. For buildings sited within Exposure B for all wind directions, the calculated pressures may be multiplied by 0.85.

4.5.5 External Pressure Coefficients for Loads on Components and Cladding, Enclosed or Partially Enclosed Buildings With Mean Roof Height Greater Than 60 ft

Values of GC_p for the design of components and cladding for enclosed or partially enclosed buildings with mean roof height greater than 60 ft are given in Fig. 6-8. Velocity pressure q_z is to be used with positive values of GC_p and velocity pressure q_h is to be used with negative values of GC_p. The reason for this distinction is that external pressures on the windward face of tall buildings are always positive and are directly related to the mean velocity pressure of the oncoming wind at height z above ground. External pressures on the side walls, leeward wall and roof of tall buildings are negative and are directly related to the mean velocity pressure at the top of the building. The pressure coefficients in Fig. 6-8 are to be used with values of q_z and q_h obtained for Exposure B, C, or D. For buildings sited in Exposure A, components and cladding are to be designed on the basis of Exposure B. This limitation reflects the uncertainty of locally higher wind speeds due to channeling effects of nearby tall buildings in built-up areas representative of Exposure A. For roof angles and geometry other than indicated in Fig. 6-8, values of GC_p from Fig. 6-5B and the attendant q_h based on Exposure C are to be used.

The values of GC_p in Fig. 6-8 have been adjusted to correspond with the new definition of basic wind speed, and the pressure zones for the walls and roof in Fig. 6-8 are simplified compared with the requirements of ASCE 7-93. The walls are now covered by two pressure zones with the interior zone extending to the top of the wall. Roof zoning has been reduced from 4 to 3 zones, and the effect of area averaging is less pronounced when compared with the requirements of ASCE 7-93. In general, the loads resulting from these pressure coefficients are lower than the loads required by ASCE 7-93. The coefficients in Fig. 6-8 were obtained from tests on isolated "box-like" buildings (Akins and Cermak 1975, Peterka and Cermak 1974) and reflect the availability of recent comprehensive wind tunnel measurements for buildings of intermediate height (Stathopoulos and Dumitrescu-Brulotte 1989).

4.5.6 Internal Pressure Coefficients for Buildings

Internal pressure coefficients, GC_{pi}, for the design of buildings are listed in Table 6-4. Four conditions are considered: 1) open buildings, 2) partially enclosed buildings, 3) buildings sited in hurricane-prone regions and having openings susceptible to damage by wind-borne debris, and 4) all other buildings. Depending upon the size and distribution of the openings, GC_{pi} may be positive or negative, and the Standard requires that each possibility be considered to establish the critical load requirements. The values of GC_{pi} are to be used in the determination of design wind loads for MWFRS and for C&C as specified in Table 6-1. When the building is sited within Exposure B for all wind directions and the mean roof height is less than or equal to 60 ft, the calculated internal pressures may be multiplied by the factor 0.85. This reduction reflects the general shielding experienced by low-rise buildings sited in rougher exposures.

The third classification in Table 6-4 recognizes the fact that glazed openings in buildings sited in hurricane-prone areas are highly susceptible to breakage by missiles or other wind-borne debris. Unless the glazing up to a height of 60 ft is specifically designed to resist missiles, or is protected by suitable shutters, it is assumed that glazed openings will be breached during an extreme wind event, resulting in high internal pressures (Minor and Behr 1993). In other words, an enclosed building will become the equivalent of a partially enclosed building unless provision has been made to prevent breaching of the glazed openings.

The fourth classification in Table 6-4 includes fully enclosed buildings and all other buildings not covered by the first three classifications. The values of GC_{pi} listed for this classification are equivalent to the old values of GC_{pi} given in ASCE 7-93 (designated Condition I) multiplied by the square of the ratio of fastest-mile speed to the 3-second gust speed. It can be seen that this same adjustment for basic wind speed does not apply when comparing the second and third classifications in Table 6-4 with what was designated as Condition II in ASCE 7-93. In fact, the internal pressures required by ASCE 7-95 represent a significant increase over those of ASCE 7-93 for this classification of openings. Supporting these new values of GC_{pi} are the results of extensive full-scale tests conducted with isolated openings representing a range of percentages of open area to gross wall area (Yeatts and Mehta 1993) and collaborating wind tunnel model studies (Yeatts et al. 1994).

4.5.7 External Pressure Coefficients for Arched Roofs

The external pressure coefficients, C_p, listed in Table 6-5 are unchanged from pervious editions of the Standard and have been established on the basis of wind tunnel model studies conducted in a uniform air stream of low turbulence. Accordingly, these coefficients are intended for use with the procedure described in Table 6-1 and in Fig. 6-3 for the design of MWFRS for enclosed or partially enclosed buildings of all heights with arched roofs. The values of C_p are dependent upon the rise-to-span-ratio, r, and the location along the external surface of the arch, either windward quarter, center half, or leeward quarter. The pressure coefficients listed in Table 6-5 are to be used when the wind is directed normal to the axis of the arch. For wind directed parallel to the axis of the arch, the roof pressure coefficients listed in Fig. 6-3 for wind parallel to the ridge are to be used.

For the design of components and cladding located in the perimeter zone of an arched roof, the values of GC_p for a gabled roof in Fig. 6-5B are to be used with the roof angle, θ, set equal to the spring-line slope of the arch. External pressure coefficients, C_p from Table 6-5, multiplied by the factor 0.87, are to be used for all other roof areas. The velocity pressure at mean roof height, q_h, used in the calculation of pressures acting on roof components and cladding are to be based on Exposure C.

4.5.8 Force Coefficients, Tables 6-6 to 6-10

With the exception of Table 6-10, the force coefficients, C_f, listed in these tables are unchanged from ASCE 7-93 and are to be applied as indicated in Table 6-1. Force coefficients for trussed towers may be calculated from the expressions listed in Table 6-10. Two basic tower cross-sections are addressed, square and triangular. The parameter ε, used in the expressions for C_f, is defined as the ratio of solid area to gross area of one tower face for the tower segment under consideration. The force coefficient is to be used with the area A_f, defined as the solid area of <u>one tower face</u> projected on the plane of that face for the tower segment under consideration.

In determining C_f, it is assumed that the tower members are structural angles or similar flat-sided shapes. For towers containing rounded members, it is acceptable to multiply the specified force coefficients by the factor $0.51\ \varepsilon^2 + 0.57$, but this factor shall not be greater than 1.0. For towers of square cross-section, wind forces shall be multiplied by the following factor when the wind is directed along a tower diagonal:

$$1 + 0.75\ \varepsilon, \text{ but not} > 1.2$$

The Standard requires that wind forces due to tower appurtenances such as ladders, conduits, lights, elevators, etc., be calculated using the appropriate force coefficients (ASCE 1961). And finally, it may be necessary to consider loads due to ice accretion as described in Section 10 of the Standard, Ice Loads--Atmospheric Icing. The force coefficients in Table 6-10 are provided as a simplified procedure for trussed towers and are consistent with force coefficients given in ANSI/EIA/TIA-222-E-1991. In a departure from previous editions of the Standard, force coefficients for the design of tower guys are no longer provided. As noted in Section 6.4.2.1 of the Standard, the designer should refer to recognized literature for documentation of wind load effects on these special structures.

4.6 Full and Partial Loading of MWFRS, Buildings With Mean Roof Height Greater Than 60 ft

The main wind-force resisting system for buildings with mean roof height greater than 60 ft shall be designed to resist the four loading cases indicated in Fig. 6-9. The loading combinations include full design wind pressures acting on the projected area perpendicular to each principal axis of the structure, partial loadings acting on two faces simultaneously, and partial loadings which induce torsional moments (twist) about the vertical axis. P_W and P_L denote the design wind pressures for the windward and leeward faces, respectively, as determined in accordance with the provisions of Table 6-1. Note that the diagrams shown in Fig. 6-9 are plan views of the building. Additional information on torsional response due to full and partial loading can be found in the literature (Isyumov 1982).

4.7 WIND TUNNEL PROCEDURE

For those situations where the analytical procedure is considered to be uncertain or inadequate, consideration should be given to wind tunnel tests. The Standard lists a set of conditions that must be satisfied for the proper conduct of such tests. The wind tunnel is particularly useful for obtaining detailed information about pressure distributions on complex shapes and the dynamic response of structures. Model scales for structural applications can range from 1:50 for a single-family dwelling to 1:400 for tall buildings. Even smaller scales may be used to model long-span bridges. Of equal importance is the ability to model complex topography at scales of the order of 1:10,000 and assess the effects of features such as hills, mountains or river gorges on the near-surface winds. Details on wind tunnel modeling for structural or civil engineering applications may be found in the literature (ASCE 1987, Cermak 1977, Reinhold 1982).

4.8 SOME FREQUENTLY ASKED QUESTIONS

1. When is a gable truss in a house part of the MWFRS ? Should it also be designed as a C&C ? What about individual members of a truss ?

   Roof trusses are considered to be components since they receive load directly from the cladding. However, trusses may be designed as part of the MWFRS if they serve a tributary area greater than 700 sq ft (Table 6-1). Even when designed as part of the MWFRS, the top chord of a gable truss may receive load directly from the roof sheathing and this should be checked using the rules for C&C. When designing shear walls or cross-bracing for light-frame construction such as a house, roof trusses can be considered as MWFRS.

2. For the design of a 4 ft high parapet, what pressure coefficients from the standard can be used for the parapet ?

   The Standard does not offer specific guidance for this situation. However, it would be appropriate to consider the parapet as a component, applying the same exterior pressure as applied to the building walls. On the interior face of the windward parapet, it would be appropriate to use GC_p for roof Zone 2. On the interior face of the remainder of the parapet, a conservative approach would be to apply the positive external pressure that acts on the windward wall of the building (See worked example 1).

3. For the determination of MWFRS loading, what pressure coefficients should be used underneath the roof overhangs and balconies ?

The MWFRS for windward roof overhangs shall be designed for a positive pressure on the bottom surface corresponding to $C_p = 0.8$ in combination with the pressures indicated in Figs. 6-3 and 6-4 (6.7.2.1). No specific guidance is given for balconies, but use of the loading criteria for roof overhangs should be adequate.

4. What constitutes an open building ? If a process plant has a three-story frame with no walls but with a lot of equipment inside the frame, is this an open building ?

An open building is a structure having all walls at least 80% open (6.2). Yes, this three-story frame could be classified as an open building or other structure in Table 6-1. In calculating the wind force, F, appropriate values of C_f and A_f would have to be assigned to the frame and to the equipment inside.

5. If the mean roof height h is greater than 60 ft with roof geometry represented by Figs. 6.5B, 6.6 or 6.7, can the pressure coefficients of these figures be used for roof component and cladding design loads ?

Yes, provided the mean roof height is less than 90 ft, the height-to-width ratio does not exceed 1, and q is taken as q_h for Exposure C (Table 6-1).

6. A building with mean roof height h of 50 ft is located in Exposure B terrain. Why is it required to use the effective velocity pressure q_h associated with Exposure C for the determination of (GC_p) from Figs. 6-4 through 6-7 ?

If the mean roof height is less than or equal to 60 ft, the building is classified as a low-rise building. If the pressure coefficients of Figs. 6-4 through 6-7 are to be used, they must be used in conjunction with q_h for Exposure C, regardless of the actual exposure of the building site. The reason for this requirement is that the pressure coefficients have been determined on the basis of Exposure C. However, the pressures so determined for the MWFRS and for C&C may be multiplied by the factor 0.85 if the building is sited in Exposure B for all wind directions (6.5.3.2.2 & 6.5.3.3.1).

7. If high winds are accompanied by rain, will the presence of raindrops increase the mean density of the air to the point where the wind loads are affected ?

Although raindrops will increase the mean density of the air, the increase is small and may be neglected. For example, if the average rate of rainfall is 5 inches per hour, the presence of raindrops will increase the mean air density by less than one percent.

8. A masonry wall is 12 ft high and 80 ft long. It is supported at the top and at the bottom. What effective wind area should be used in determining the design pressure for the wall ?

For a given condition, the magnitude of the pressure coefficient, GC_p, increases with decreasing effective wind area. Therefore, a conservative approach would be to consider an effective wind area with a span of 12 ft and a width of 1 ft, and design the wall element as C&C. However, the definition of effective wind area states that this area is the span length multiplied by an effective width that need not be less than one-third the span length. Accordingly, the effective wind area would be (12)(12/3) = 48 sq ft.

9. Roof trusses are 30 ft long and are spaced on 4 ft centers. What effective wind area should be used to determine the design pressures for the trusses ?

Roof trusses are classified as components and cladding since they receive wind load directly from the cladding (roof sheathing). In this case the effective wind area is the span length multiplied by an effective width that need not be less than one-third the span length or (30)(30/3) = 300 sq ft. This is the area on which the selection of GC_p should be based. However, the resulting wind pressure acts on the tributary area of each truss which is (30)(4) = 120 sq ft.

10. Roof trusses have a clear span of 70 ft and are spaced 8 ft on center. What effective wind area should be used to determine the design pressures for the trusses ?

Following the approach of the previous example, the effective wind area is (70)(70/3) = 1633 sq ft. The tributary area of the truss is (70)(8) = 560 sq ft which is less than the 700 sq ft area required by Table 6-1 to qualify for design of the truss using the rules for MWFRS. The truss is to be designed using the rules for C&C, and the wind pressure corresponding to an effective wind area of 1633 sq ft is to be applied to the tributary area of 560 sq ft.

11. Metal decking consisting of panels 20 ft long and 2 ft wide is supported on purlins spaced 5 ft apart. Will the effective wind area be 40 sq ft for the determination of pressure coefficients ?

Although the length of a decking panel is 20 ft, the basic span is 5 ft. According to the definition of effective wind area, this area is the span length multiplied by an effective width that need not be less than one-third the span length. This gives a minimum effective wind area of (5)(5/3) = 8.3 sq ft. However, the actual width of a panel is 2 ft, making the effective wind area equal to the tributary area of a single panel or (5)(2) = 10 sq ft. Therefore, GC_p would be determined on the basis of 10 sq ft effective wind area, and the corresponding wind load would be applied to a tributary area of 10 sq ft. Note that GC_p is constant for effective wind areas less than 10 sq ft.

12. A tower has a fundamental frequency of 2 Hz, but its height to width ratio is 6. Should the tower be treated as a flexible structure to determine the gust effect factor ?

No. The guideline of height-to-width ratio of 4 or more is intended to save the user of the Standard the trouble of calculating the fundamental frequency in each and every case. The energy in the turbulence spectrum is very small for frequencies above 1 Hz. Hence, a tower with fundamental frequency of 2 Hz. will not be dynamically excited.

13. If the design wind loads are to be determined for a building that is located in a special wind region (shaded area in Fig. 6-1), what basic wind speed should be used ?

The purpose of the special wind regions in Fig. 6-1 is to alert the user to the fact that there are regions in which wind speed anomalies are known to exist. Wind speeds in these regions may be substantially higher than the speeds indicated on the map, and the use of regional climatic data and consultations with a wind engineer or consulting meteorologist are advised.

14. If a monoslope roof over an open building is virtually flat, what force coefficients from Table 6-6 should be used ?

A requirement for the use of Table 6-6 is that the wind shall be assumed to deviate ± 10 degrees from the horizontal. Accordingly, the values of C_f corresponding to a roof angle of 10 degrees should be used. The wind forces may be directed either inward or outward, and both cases should be checked.

15. A trussed tower of 10 x 10 ft square cross-section consists of structural angles forming basic tower panels 10 ft high. The solid area of the face of one tower panel projected on the plane of that face is 22 sq ft. What force coefficient, C_f, should be used to calculate the wind force ? What would the force coefficient be for the same tower fabricated of rounded members having the same projected solid area ? What area should be used to obtain the wind force per ft of tower height acting (1) normal to a tower face, and (2) along a tower diagonal ?

The gross area of one panel face is (10)(10) = 100 sq ft, and the solidity ratio is $\varepsilon = 22/100 = 0.22$. For a tower of square cross-section, the force coefficient is as follows:

$$\begin{aligned} C_f &= (4)(0.22)^2 - (5.9)(0.22) + 4.0 \\ &= 2.90 \end{aligned}$$

For rounded members, the force coefficient may be reduced by the factor

$$(0.51)(0.22)^2 + 0.57 = 0.59$$

Thus the force coefficient for the same tower constructed of rounded members with the same projected area would be

$$\begin{aligned} C_f &= (0.59)(2.90) \\ &= 1.71 \end{aligned}$$

The area, A_f, used to calculate the wind force per ft of tower height is 22/10 = 2.2 sq ft for all wind directions.

16. Eq. 6-1 for velocity pressure uses the subscript z while Tables 6-1 and 6-3 use subscripts z and h. When is z used and when is h used ?

Eq. 6-1 is the general formula for the velocity pressure q_z at any height z above ground. There are many situations in the Standard where a specific value of z is called for, namely the height (or mean roof height) of a building or other structure as indicated in Table 6-1. Whenever the subscript h is called for in Table 6-1, it is understood that z becomes h in Eq. 6-1 and in Table 6-3.

17. The Standard does not provide for across-wind excitation caused by vortex shedding. How can one determine when vortex shedding might become a problem?

Vortex shedding is almost always present with bluff-shaped cylindrical bodies. It can become a problem when the frequency of shedding is close to or equal to the frequency of the 1st or 2nd transverse modes of the structure. The intensity of excitation increases with aspect ratio (height-to-width or length-to-breadth) and decreases with increasing structural damping. Structures with low damping and with aspect ratio of 8 or more may be prone to damaging vortex excitation. If across-wind excitation appears to be a possibility, expert advice should be obtained.

18. Under what conditions is it necessary to consider topographic effects (speed-up) when calculating wind loads?

The Standard requires the calculation of the topographic factor K_{zt} for buildings and other structures sited on the upper half of isolated hills or escarpments located in Exposure B, C, or D where the upwind terrain is free of such topographic features for a distance of at least 50H or one mile, whichever is smaller, as measured from the crest of the topographic feature. K_{zt} need not be calculated when the height H is less than 15 ft in Exposure D, or less than 30 ft in Exposure C, or less than 60 ft in Exposure B. In addition, K_{zt} need not be calculated when H/L_h is less than 0.2. H and L_h are defined in Fig. 6-2. The value of K_{zt} is never less than 1.0.

19. Fig. 6-9 shows wind pressure combinations to be used for full and partial loading in the design of main wind-force resisting systems in buildings with mean roof height greater than 60 ft. Does the Standard have similar requirements for buildings with mean roof height less than or equal to 60 ft?

No. The Standard does not require the designer to consider the partial loadings of Fig. 6-9 when designing buildings with mean roof height less than or equal to 60 ft. However, the actions of diagonally-acting resultants and torsional moments are accounted for automatically when using the external pressure coefficients, GC_{pf}, given for load cases A and B in Fig. 6-4.

20. In the design of main wind-force resisting systems, the provisions of Fig. 6-3 apply to enclosed or partially enclosed buildings of all heights. The provisions of Fig. 6-4 apply to enclosed or partially enclosed buildings with mean roof height less than or equal to 60 ft. Does this mean that either figure may be used for the design of low-rise MWFRS?

Yes. The choice of approach is left to the designer. Fig. 6-3 allows the use of velocity pressures based on Exposure A, B, C or D. Fig. 6-4 requires the use of velocity pressure, q_h, associated with Exposure C, but allows the wind loads so determined to be reduced by the factor 0.85 if the building is sited in Exposure B for all wind directions.

21. In calculating the wind forces acting on a trussed tower of square cross section (Table 6-10), should the force coefficient, C_f, be applied to both the front and the back (windward and leeward) faces of the tower?

No. The calculated wind forces are the total forces acting on the tower. The force coefficients given in Table 6-10 include the contributions of both front and back faces of the tower as well as the shielding effect of the front face on the back face.

22. Is it possible to obtain a larger scale map of basic wind speeds (Fig. 6-1) so that the locations of the wind speed contours can be determined with greater accuracy?

The wind speed contours in the hurricane-prone region of the United States are based on hurricane wind speeds derived from Monte Carlo simulations and on estimates of the rate at which hurricane wind speeds reduce to 90 mph following landfall. Because the wind speed contours of Fig 6-1 represent a consensus of the ASCE 7 Wind Load Task Committee, a larger scale map would do nothing to improve their accuracy.

23. If the pressure or force coefficients for various roof shapes, e.g. a canopy, are not given in ASCE 7-95, how can the appropriate wind forces be determined for these shapes?

With the exception of pressure or force coefficients for certain shapes, parameters such as V, I, K_z, K_{zt} and G are given in ASCE 7-95. It is possible to use pressure or force coefficients from the published literature (see Section 1.4) provided these coefficients are used with care. Mean pressure or force coefficients from other sources can be used to determine wind loads for MWFRS. However, it should be recognized that these coefficients may have been obtained in wind tunnels having smooth, uniform flows as opposed to more proper turbulent boundary layer flows. Pressure coefficients for components and cladding obtained from the literature should be adjusted to the 3-second gust speed which is the basic wind speed adopted by ASCE 7-95.

24. Can the pressure/force coefficients given in ASCE 7-95 be used with the provisions of ASCE 7-88 or 7-93?

Yes, in a limited way. ASCE 7-88 (and 7-93) used the fastest-mile wind speed as the basic wind speed. With the adoption of the 3-second gust speed in ASCE 7-95, the values of certain parameters used in the determination of wind forces have been changed accordingly. The provisions of ASCE 7-88 and 7-95 should not be interchanged.

LIST OF REFERENCES

Akins, R.E. and Cermak, J.E. (1975). "Wind pressures on buildings," Technical Report CER7677REA-JEC15, Fluid Dynamics and Diffusion Laboratory, Colorado State University, Fort Collins, CO.

ANSI A58.1-1982 (1982). "Minimum Design Loads for Buildings and Other Structures," American National Standards Institute, New York, New York.

ANSI/EIA/TIA-222-E (1991). "Structural standards for steel antenna towers and antenna supporting structures," Electronic Industries Assn., Washington, DC.

ASCE (1961). "Wind forces on structures," *Transactions,* ASCE, 126 (2), 1124-1198.

ASCE (1987). "Wind tunnel model studies of buildings and structures," *Manuals and Reports on Engineering Practice,* No. 67, American Society of Civil Engineers, New York, NY.

ASCE/ANSI 7-88 (1990). "Minimum Design Loads for Buildings and Other Structures," American Society of Civil Engineers, New York, New York.

ASCE/ANSI 7-95 (1996). "Minimum Design Loads for Buildings and Other Structures," American Society of Civil Engineers, New York, New York.

AS-1170.2 (1989). "Australian Standard SAA Loading Code Part 2," Wind Loads, Standards Australia

Batts, M.E., Cordes, M.R., Russell, L.R., Shaver, J.R. and Simiu, E. (1980). "Hurricane wind speeds in the United States," NBS Building Science Series 124, National Bureau of Standards, Washington, DC.

Best, R.J. and Holmes, J.D. (1978). "Model study of wind pressures on an isolated single-story house," Wind Engineering Report 3/78, James Cook University of North Queensland, Australia.

British Standard/BS 6399 (1995). "Loading for Buildings, Part 2: Code of Practice for Wind Loads," British Standards Institute, London W4 4AL

Cermak, J.E. (1977). "Wind-tunnel testing of structures," *Journal of Engineering Mechanics Division.,* ASCE, 103(6), 1125-1140.

Cook, N.J. (1985). The Designer's Guide to Wind Loading of Building Structures, Parts I and II, Butterworth Publishers, London, England.

Davenport, A.G., Surry, D. and Stathopoulos, T. (1977). "Wind loads on low-rise buildings," Final Report on Phases I and II, BLWT-SS8, University of Western Ontario, Canada.

Davenport, A.G., Surry, D. and Stathopoulos, T. (1978). "Wind loads on low-rise buildings," Final Report on Phase III, BLWT-SS4, University of Western Ontario, Canada.

Durst, C.S. (1960). "Wind speeds over short periods of time," *Meteorological Magazine,* 89, 181-187.

Eaton, K.J. and Mayne, J.R. (1975). "The measurement of wind pressures on two-story houses at Aylesbury," *Journal of Industrial Aerodynamics,* 1(1), 67-109.

EUROCODE 1 (1994). *Basis of Design and Actions on Structures/Wind Action, Part 2.3: Wind Actions,* CEN/TC 250/SC1, Steering Panel Draft, May 1994.

FEMA (1980). *Interim Guidelines for Building Occupant Protection from Tornadoes and Extreme Winds*, TR83-A, Federal Emergency Management Agency, Washington, DC.

Georgiou, P.N., Davenport, A.G. and Vickery, B.J. (1983). "Design wind speeds in regions dominated by tropical cyclones," *Journal of Wind Engineering and Industrial Aerodynamics,* 13, 139-152.

Ho, E. (1992). "Variability of low building wind loads," Doctoral Dissertation, University of Western Ontario, London, Ontario, Canada.

Holmes, J.D., Melbourne, W.H., and Walker, G.R. (1990). "A Commentary on the Australian Standard for Wind Loads," Australian Wind Engineering Society (printed by Courtney Colour Graphics Pty. Ltd., Lilydale, Victoria, Australia).

Hoerner, S.F.(1965). Fluid Dynamics Drag, published by the author, Midland Park, NJ.

Isyumov, N. (1982). "The aeroelastic modeling of tall buildings," *Proceedings,* International Workshop on Wind Tunnel Modeling Criteria and Techniques in Civil Engineering Applications, NBS, Gaithersburg, MD, Cambridge University Press, 373-407.

Isyumov, N. and Case, P. (1995). "Evaluation of structural wind loads for low-rise buildings contained in ASCE Standard 7-1995," BLWT-SS17-1995, University of Western Ontario, London, Ontario, Canada.

Jackson, P.S. and Hunt, J.C.R. (1975). "Turbulent wind flow over a low hill," *Quarterly Journal of the Royal Meteorological Society,* 101, 929-955.

Kareem, A. (1992). "Dynamic response of high-rise buildings to stochastic wind loads," *Journal of Wind Engineering and Industrial Aerodynamics,* 41-44.

Kareem, A. (1985). "Lateral-torsional motion of tall buildings to wind loads," *Journal of Structural. Engineering,* ASCE 111(11)

Kareem, A. and Smith, C. (1993). "Performance of offshore platforms in Hurricane Andrew," *Proceedings,* Hurricanes of 1992, ASCE, Dec. 1-3, Miami, FL, Dec.

Kavanagh, K.T., Surry, D., Stathopoulos, T. and Davenport, A.G. (1983). "Wind loads on low-rise buildings: Phase IV," BLWT-SS14, University of Western Ontario, London, Ontario, Canada.

Krayer, W.R. and Marshall, R.D. (1992). "Gust factors applied to hurricane winds," *Bulletin of the American Meteorological Society,* Vol. 73, 613-617.

Lawson, T.V. (1980). Wind Effects on Buildings, Volumes 1 and 2, Applied Science Publishers Ltd., Ripple Road, Barking, Essex, England

Lemelin, D.R., Surry, D. and Davenport, A.G. (1988). "Simple approximations for wind speed-up over hills," *Journal Wind Engineering and Industrial Aerodynamics,* 28, 117-127.

Liu, Henry (1991). Wind Engineering: A Handbook for Structural Engineers, Prentice-Hall, New York, NY.

Marshall, R.D. and Yokel, F.Y. (1995). Recommended performance-based criteria for the design of manufactured home foundation systems to resist wind and seismic loads, NISTIR 5664, National Institute of Standards and Technology, Gaithersburg, MD.

McDonald, J.R. (1983). "A Methodology for Tornado Hazard Probability Assessment," NUREG/CR3058, U.S. Nuclear Regulatory Commission, Washington, DC.

Mehta, K.C., Minor, J.E. and McDonald, J.R. (1976). "Wind speed analyses of April 3-4 tornadoes," *Journal of Structural Division,* ASCE, 102(9), 1709-1724.

Minor, J.E. (1982). "Tornado technology and professional practice," *Journal of Structural Division,* ASCE, 108(11), 2411-2422.

Minor, J.E. and Behr, R.A. (1993). "Improving the performance of architectural glazing systems in hurricanes," *Proceedings,* Hurricanes of 1992, ASCE, Dec. 1-3, Miami, FL, pp C1-11.

Minor, J.E., McDonald, J.R. and Mehta, K.C. (1977). The tornado: An engineering oriented perspective, TM ERL NSSL-82, National Oceanic and Atmospheric Administration, Environmental Research Laboratories, Boulder, CO.

Murray, R.C. and McDonald, J.R. (1993). Design for containment of hazardous materials, *Geophysical Monograph 79,* The Tornado: Its Structure, Dynamics, Prediction and Hazards, C. Church, D. Burgess, C. Doswell and R. Davies-Jones, Eds, American Geophysical Union, 379-387.

Newberry, C. W., and Eaton, K.J. (1974). *Wind Loading Handbook*, Building Research Establishment Report K4F, Her Majesty's Stationery Office, London, England.

NRCC (1995a). *National Building Code of Canada, 1995*, Associate Committee on the National Building Code of Canada, National Research Council of Canada.

NRCC (1995b). *Supplement to the National Building Code of Canada, 1995*, Associate Committee on the National Building Code of Canada, National Research Council of Canada.

Peterka, J.A. (1992). "Improved extreme wind prediction for the United States," *Journal of Wind Engineering and Industrial Aerodynamics,* 41, 533-541.

Peterka, J.A. and Cermak, J.E. (1974). "Wind pressures on buildings--Probability densities," *Journal of Structural Division,* ASCE, 101(6), 1255-1267.

Peterka, J.A. and Shahid, S. (1993). "Extreme gust wind speeds in the U.S.," *Proceedings,* 7th U.S. National Conference on Wind Engineering, UCLA, Los Angeles, CA, 2, 503-512.

Reinhold, T.A. (Ed.) (1982). "Wind tunnel modeling for civil engineering applications," *Proceedings,* International Workshop on Wind Tunnel Modeling Criteria and Techniques in Civil Engineering Applications, NBS, Gaithersburg, MD, Cambridge University Press.

SAA (1989). *Australian Standard SAA Loading Code, Part 2: Wind Loads*, published by Standards Australia, Standards House, 80 Arthur St., North Sydney, NSW, Australia.

Saathoff, P. and Stathopoulos, T. (1992). "Wind loads on buildings with sawtooth roofs," *Journal of Structural Engineering,* ASCE, 118(2), 429-446.

Simiu, E. and Scanlan, R.H. (1996). Wind effects on structures, Third Edition, John Wiley & Sons, New York, NY.

Solari, G. (1993). "Gust buffeting I: Peak wind velocity and equivalent pressure," *Journal of Structural Engineering,* ASCE, 119(2).

Solari, G. (1993). "Gust buffeting II: Dynamic along-wind response," *Journal of Structural Engineering,* ASCE, 119(2).

Stathopoulos, T. (1981). "Wind loads on eaves of low buildings," *Journal of Structural Division,* ASCE, 107(10), 1921-1934.

Stathopoulos, T. and Dumitrescu-Brulotte, M. (1989). "Design recommendations for wind loading on buildings of intermediate height," *Canadian Journal of Civil Engineering,* 16(6), 910-916.

Stathopoulos, T. and Luchian, H.D. (1990). "Wind pressures on building configurations with stepped roofs," *Canadian Journal of Civil Engineering,* 17(4), 569-577.

Stathopoulos, T. and Mohammadian, A.R. (1986). "Wind loads on low buildings with mono-sloped roofs," *Journal of Wind Engineering and Industrial Aerodynamics,* 23, 81-97.

Stathopoulos, T. and Saathoff, P. (1991). "Wind pressures on roofs of various geometries," *Journal of Wind Engineering and Industrial Aerodynamics,* 38, 273-284.

Surry, D. and Stathopoulos, T. (1988). "The wind loading of buildings with monoslope roofs," Final Report, BLWT-SS38, University of Western Ontario, London, Ontario, Canada.

Vickery, P.J. and Twisdale, L.S. (1993). "Prediction of hurricane wind speeds in the U.S.," *Proceedings,* 7th U.S. National Conference on Wind Engineering, UCLA, Los Angeles, CA, 2, 823-832.

Walmsley, J.L., Taylor, P.A. and Keith, T. (1986). "A simple model of neutrally stratified boundary-layer flow over complex terrain with surface roughness modulations," *Boundary-Layer Meteorology,* 36, 157-186.

Wen, Y.K. and Chu, S.L. (1973). "Tornado risks and design wind speed," *Journal of the Structural Division,* ASCE, 99(12), 2409-2421.

Yeatts, B.B. and Mehta, K.C. (1993). "Field study of internal pressures," *Proceedings,* 7th U.S. National Conference on Wind Engineering, UCLA, Los Angeles, CA, 2, 889-897.

Yeatts, B.B., Womble, J.A., Mehta, K.C. and Cermak, J.E. (1994). "Internal pressures for low-rise buildings," *Proceedings,* Second U.K. Conference on Wind Engineering, Warwick, England.

SUBJECT INDEX